Additional Praise for Climate Dragon

"Sandy Lawrence weaves a story employing his fictional characters of the climate change tempest facing our planet. The challenges presented by climate science, nuclear energy, and a bit of healthcare science highlight the present and future battles our civilization must address in this well-written tale."

—Barry M. Meyers, Certified Emeritus Elder Law Attorney, Bellingham, Washington

"*Climate Dragon* brings both humor and humanity to a suspenseful storyline set against a backdrop of climate chaos and clean energy progress. You'll learn a ton about climate and energy—and you'll want to keep turning the pages."

—Lisa Hymas, Executive Editor at Canary Media, Seattle, Washington

"Kudos to this physician author for his first novel which is akin to a tapestry woven from a combination of high-level science and fiction. (A bit like adding the medicine prescribed into a dessert, so the patient will consume it.) It might be a bit more erudite, scientific, and pedantic than the average consumer will understand or appreciate, however. Overall, I find it a fascinating and novel read given the author's keen ability to incorporate so much factual science into his first attempt at literary fiction."

—Jonathan Franklin, MD, Neuroradiologist, retired senior member American College of Radiology, retired founding member Northwest Radiologists, Bellingham, Washington

"If you are curious about the world's changing climate and the valiant efforts to design viable solutions to our global energy challenges, check out *Climate Dragon*—an engaging tale of exploration and discovery."

—David Roberts, President, Peak Sustainability Group, Bellingham, Washington

"I'm proud of my friend for taking such a creative approach to address the complex topics of climate change. His efforts to open minds with *Climate Dragon* is most admirable. The story addresses the impacts of environmental change that affect not just the biological health of the planet, but the political, social, and economic consequences for its inhabitants. The blend of factual information and the fictional characters presents an effective way for us to make valuable differences in the way we live. Sometimes seriously dramatic, sometimes comically lighthearted, the book peels away at the prospects for our future. It is my hope that readers will be inspired to start meaningful and impactful conversations around climate issues."

—Jerry Thon, Education degree at Western Washington University, Former President of the Foundation for WWU and Alumni Former fishing representative for the Pacific Coast Management Council in the National Oceanographic and Atmospheric Administration

"*Climate Dragon* is a must read for anyone concerned with the future of our planet. In this story by Dr. Lawrence, Jack and Abbey take center stage in an illuminating exploration of renewable energy. Lawrence skillfully intertwines a compelling narrative with real-world insights, making complex concepts accessible to readers of all ages."

—Lorinda Boyer, author, Straight Enough

CLIMATE DRAGON

CLIMATE DRAGON
TREACHERY, PESTILENCE & WEIRDING WEATHER

S.W. Lawrence, MD

Sidekick Press
Bellingham, Washington

Publisher's Note: This is a work of fiction. Names, characters, places, and incidents are products of the author's imagination. Locales and public names are sometimes used for atmospheric purposes. Any resemblances to actual people, living or dead, or to businesses, companies, events, institutions, or locales is completely coincidental.

Published 2024
Printed in the United States of America
ISBN: 978-1-958808-17-7
LCCN: 2023920187

Sidekick Press
2950 Newmarket Street, Suite 101-329
Bellingham, Washington 98226
sidekickpress.com

Climate Dragon: Treachery, Pestilence, and Weirding Weather

Cover design by Bob Paltrow
Cover image: "Prismatic Dragon Head" by Gordan Dylan Johnson
Half-title image: Artist unknown

Quotes Used by Permission

For my Barbara

and the whole Schickler clan

For walk where we will, we tread upon some story.
—Marcus Tullius Cicero

CONTENTS

Planetary Odyssey

He who returns from a journey is not the same as he who left.
—*Chinese proverb*

Voyage, travel, and change of place impart vigor.
—*Seneca*

The first step of a journey may be taken with either resolution or irresolution, but either one constitutes a commitment. The only certitude is that one may not be the same person by the end. It is my fond hope this manuscript may alter and broaden people's outlooks, and encourage them to cultivate better understanding and concern for our present planetary dilemma.

Now thundering into our coastlines is a succession of massive breakers of crushing intensity—the manifestations of outsized storms, rising sea levels, marine acidification, deoxygenation, finfish and kelp extinctions, and all the other components of the weirding of our seas and wider world.

We are buoyed by our appreciation of the potentials of energy efficiency and a smart grid, wind and solar power, tidal power and hydropower, and other renewable energy flows supporting a sustainable future.

Radical optimism remains necessary, because the plentitude of dysfunctions in our ecosystem is daunting.

Our tale of derring-do will explore novel dilemmas. One stems from the unique vulnerabilities of our nuclear legacy. Another derives from the migrations of pestilence as tropical diseases spread.

I hope you're as curious as I am to see how this all plays out.

ONE

Dragon Teeth

NCCIC delivers a full spectrum of cyber exercise planning workshops and seminars, and conducts tabletop, full-scale, and functional exercises, as well as the biennial National Cyber Exercise . . . designed to assist organizations at all levels in the development and testing of cybersecurity prevention, protection, mitigation, and response capabilities.
—US Department of Homeland Security

His fingers were long, genuinely spidery. With analytic detachment, he watched his index finger lightly tap and then, finally, press enter on the keyboard, albeit with a slight tremor—whether from anticipation or apprehension or caffeine, it wasn't quite clear. The ironic thought struck him that this moment's outcome was just "peachy keen." He chuckled at his own joke, knowing his months of preparation had culminated in . . . he frowned in search of the right word . . . *fruition*, so to speak, then slowly grinned at his compounded jest. Finished finally with a deep, cleansing breath.

The actual target lay only an hour away as the crow flies, but he had decided superstitiously to forgo direct reconnaissance. His Dragon had pinballed through multitudinous cities and server computers around the world before tiptoeing inside and coming to rest, to slumber with dreams of electric sheep. The beast need not awaken for months.

Three glowing screens lit the sparse and spacious attic room. The desks were spotless, as was the entirety of the room, since he wouldn't tolerate spiderwebs. But an ineradicable, musty-mold smell was embedded in the walls from roof leaks the landlord had procrastinated in fixing. A collector of maps since childhood, he had mounted his favorite specimens on several walls, with pride of place given to an ancient map of the English colonies in North America, discovered in an antique shop in Vermont years ago.

Piles of notes and scrap paper and a container of pens were carefully arranged on the upper right corner of an ancient wooden desk—which was really two desks joined end to end, one of them blocked up for the purpose of leveling. On the left desk stood a pair of antique sand timers, whose accuracy he had confirmed with the old but still functional satellite clock above one of the windows. He was most happy while working in beatific solitude. Quiet was imperative, as even a ticking clock set his teeth on edge.

He reached down to pet his dog, a pit bull Akita mix, a rescue. Someone had surrendered the animal to the pound because of its aggressive behavior, which he found actually attractive. Occasionally he took his dog for long runs at night, never with a jaw restraint but always on a leash. The animal was protective and loyal but had growled and lunged at people after dark. The man knew the last thing he needed was attention from the authorities.

His dog was not his only friend. A cadre of multiplayer gamers admired and respected him, despite never having met him in person. They knew him only as "Muzzle Flash," and he made sure even these friends couldn't locate him on his encrypted private network.

To cover his tracks, he used cryptocurrency on the dark web to purchase his coveted hacking tools. Key was his coup in obtaining a version of the Israeli spyware, Pegasus,[1] allowing him to reliably and consistently decipher encrypted communications on iPhones or Android phones. Also critical was obtaining the companion software, Phantom,[2] a backdoor program allowing him to extend his access to American phone numbers. He sourced other critical components from authoritarian countries like Hungary and Poland. In the latter case, his fluency in Polish from childhood was instrumental.

The underlying math was solidly based on elliptic-curve[3] cryptography, which had been used to solve the famous mathematical riddle of Fermat's last theorem. He knew his principal foe was the National Security Agency, the largest of the eighteen American intelligence services. Or were their other entities buried so deeply they didn't show up on any organizational charts?

Both he and the dog startled when some late evening firecrackers exploded nearby in the run-up to the Fourth of July, always a big event in Washington, DC. Neither tolerated loud noises or confrontation well.

He grimaced at the childhood memories that plagued him at inopportune times—the pediatrician who had identified his awkwardly long fingers as arachnodactyly, though she reassured his parents he had no other genetic anomalies; the ophthalmologist who described his color-mismatched eyes as "heterochromia iridic," hazel on the right but gray or pale blue on the left, varying with ambient lighting; the orthodontist who insisted on braces until he was seventeen. No wonder the kids in high school thought he was a freak.

His pallor wasn't unhealthy; it just spoke of avoiding sunlight. He was gangly but not gaunt, with meticulously clean fingernails and no visible tattoos or piercings. His reading glasses were a constant companion, tethered around his neck, though switched out for wraparound sunglasses he wore any time he braved exposure outside. He wasn't so much sensitive to bright light as to unwelcome attention from strangers, though he appreciated young children. Not in any unwholesome way, he told himself, but owing to their disregard of his subtle differences.

He had blossomed once he survived the early, ritualistic school hazing. He'd enjoyed the loneliness of the long-distance runner, even in his teenage years, but had not tried out for any athletic team until his last year in high school. His coaches had tried to recruit him as far back as sophomore year. In college, though, he ran both cross-country and steeplechase and made some respectable showings, but dropped off both teams by the middle of his second year. By then he had succumbed to the allure of mathematics and computers and software engineering, which had not just become his majors but had tugged him into a vortex of obsession.

Another firecracker exploded. He picked up and cradled the dog in his arms, leading both to become calmer.

TWO

Ruminations

It's the place of story, beginning here, in the meadow of late summer flowers, thriving before the Atlantic storms drive wet and winter upon them all.
—*Gregory Maguire,* Confessions of an Ugly Stepsister

Jake slept fitfully, for reasons unclear, until in the early predawn hours he awoke to soft moanings and complaints from Abbey, with morning sickness her recently inaugurated, primal suffering. She berated him about his tossing and turning, unfairly he thought, but he'd committed himself to not complaining.

Eyes barely open, he got up and got dressed in the dark while sitting on the bedside chair, fumbling first with the wrong shoe, then with tangled shoelaces. He exited the room quietly, closing the door behind him with the exquisite care used in setting a mouse trap, then in the bathroom threw some cold water on his face. He looked in the mirror and saw lines of fatigue framing his nut brown eyes. He finished quickly, using his fingers on both sides to sweep back his wheat-colored hair. Only a barber had ever used a comb on him.

He resolved to make nothing more complicated for the only guests staying at the inn than scrambled eggs mixed with bits of smoked salmon. Maybe a little garnish of parsley. Coffee, of course, and cereal—several options. Anything else left over in the fridge from the weekend. Out of fresh fruit, alternatively, he could offer some raisins or dried apricots. He loaded up the coffee maker in the kitchen, fortunately having remembered to grind the beans the night before. The rest could wait, as it was not yet five thirty. The sky outside was as dark as squid ink.

He was too weary to do anything more except retrieve the two newspapers thrown at the Dragonfly Inn sign out by the street. *Can always read them later.* He decided just to grab a throw pillow and get some shut-eye on the couch in the parlor. Thoughts jumbled in his mind as he faded into oblivion.

The next thing he knew, Abbey was grabbing his arm and whispering fiercely, "Jake, damn it all, I just got up to throw up, but now I hear the guests stirring upstairs. Get your butt in gear, I have to get dressed. I told you to wake me up half an hour ago so I wouldn't be late getting to the hospital."

The smattering of freckles beside her nose seemed pale when she came out of the bedroom buttoning her blouse. Her lush hair was the color of dark garnet, tied back in a thick ponytail. She gave him a quick, tentative smile and hugged him to make up for her waking brusqueness. Then she was out the door like an autumn gust of wind.

It wouldn't be until much later, after the guests had come down for breakfast, that Jake would realize he had put on yesterday's shirt, complete with spaghetti sauce stains.

Abbey had rationalized her nausea and dry heaving as signs of elevated progesterone, consistent with sturdy placental development and, ironically, a lower risk of miscarriage.

This consolation was only an intellectual exercise as far as he was concerned. *She can't blame me for the pregnancy, can she? After all, it was her idea. Mainly.*

THREE

Lady with the Gold Dress On

That's what she was, broken pottery, patched up with gold, the gold shimmering through the places where she had been cracked open, and left bleeding.
—Kiran Manral

All bleeding eventually stops.
—Anonymous surgeon

The paramedics who arrived just after the first patrol car on the scene were appropriately dressed in light blue PPE—including N-95 masks equipped with splash guards—and followed standard Covid protocol in assessing the patient. They found themselves in an Anacostia second-floor, walk-up apartment, with empty cans and other paraphernalia littering the dusty floor.

Kristy immediately tasked Mike with getting intravenous access and starting an IV line. She saw at once that the lady breathing shallowly on the couch was bleeding everywhere, with stringy blood and mucus from the nose; diffuse, small purple petechial dots on her skin and in her eyes; even bleeding gingivae when they examined her gums and throat with a tongue blade. Her forehead temperature was 102.5 degrees Fahrenheit and her blood oxygen level was 73, with normal above 95. She moaned and turned her head from side to side.

Bleeding and fever and low oxygen—not a great combination.

Her tattered dress of goldenrod yellow was already sodden in front with blood, as was the cushion of the old, gray, swaybacked couch. A sudden, gurgling cough splattered brighter red blood on Kristy's face shield. The lead on the scene, she cursed under her breath. She reached for a quick four-by-four-inch gauze sponge to wipe the shield clean. The patient's pallor bleached out her dark skin, and her pulse was faint, thready, and tachycardiac, almost 160.

The two police officers were gloved and masked but stayed well out of the way.

The patient was minimally responsive to pain when the second paramedic, burly, sweaty Mike, struggled to start an IV, the procedure taking extra seconds because of the multiple tracks and scars on her hands and forearms. He finally got a large-bore needle into a vein in the right antecubital fossa, inside the curve of the elbow. But pushing two ampules of naloxone directly into that vein elicited no response—unsurprising, given her dilated pupils— suggesting narcotic overdose wasn't the primary problem.

Kristy had taught Mike that ninety percent of people are right-handed and hence abuse their left-sided veins more frequently, at least in the beginning. She knew of a clinical test to detect brain function on either side, simply by comparing the size of left and right thumbs. Often the dominant thumb is wider. But that fact was irrelevant. The patient was already in extremis.

Once they lifted her to the gurney and secured her in place, they rapidly loaded her into the ambulance. Mike headed out with full siren and lights for George Washington Medical Center.

Mike radioed ahead, relaying that they had a patient with presumed coagulopathy and probable hypovolemic shock bleeding out. They roared through town, scattering the dry leaves of fall after a long, hot summer. In the last minute before arrival, the patient became unresponsive and pulseless, though the EKG showed an erratic, idioventricular rhythm. But Kristy knew with bottoming-out blood volume, whatever the rhythm, the heart had no pump primer, no blood to pump forward, nothing to send to the critical tissues of heart muscle and brain.

Kristy struggled as she attempted intubation because of the amount of blood in the pharynx and airway, not suctioned adequately with ongoing hemorrhage, and again cursed angrily but quietly.

She resorted to using a face mask to bag the woman with the gold dress. The receiving crew in the ambulance bay flung back the back doors abruptly as the woman's chances of survival plummeted.

Her life was inexorably running down, grains of sand through an hour-glass.

FOUR

Daybreak Unveiling

Nothing in life is a foregone conclusion unless and until it is foregone and concluded.
—*Rasheed Ogunlaru*

Jake had always been a clutch player, able to pull out a strong kick at the end of a race. Abbey had submitted to a quick kiss on the cheek on the way out the door, still grumbling, sotto voce. Though she'd brushed her teeth, she had been kind enough to not subject Jake to her dragon breath. After he had escorted her out the door with encouragement and a pat on the rear, he got right to work, his game face on, focused, with two days' worth of dark stubble.

In minutes, he'd set the kitchen table for two with cloth napkins and matching silverware. Having been told the guests had a toddler, he brought the wooden high chair from the dining room and angled it next to one corner. He placed out of reach of a small child two offerings of fruit, including blueberries—his favorite, ripe in September—plus dried apricots. Bowls of brown sugar and raisins he set next to a beautiful orchid in a small ceramic vase. He whipped up some eggs and shredded the leftover salmon to mix in once the eggs started to coagulate in the frying pan. The oatmeal was nearing the lowest boil as the guests came down the stairs. Jake relaxed as he realized he'd gotten set up in time. *Thanks to the gods for small favors.*

He heard the baby first, burbling as she was carried downstairs. Jake poked his head out the doorway to say hello, and was immediately entranced by the sight of the toddler, dressed in a white jumpsuit emblazoned with green frogs. *Cute kid alert*, he thought to himself. "You must be Ben and Emmanuelle, or so I heard from Abbey last night. But who is this gorgeous young child?"

Ben, holding the toddler in his arms, took one last step off the stairs, then set her down to let her stand decisively on her own and introduced her as Margaux, hastening to add that they used not the English but the French spelling of her name. Focusing on meeting the parents, Jake resisted the urge to squat down to the infant's height, instead waiting for the child to advance. But she reached for her father's hand without taking a step, staring wide-eyed at Jake.

Ben reached out to shake hands with him, smiling proudly as he did so. Jake noted Ben's prominent nose and ears, dark hair parted on the left, a light gray, long-sleeved shirt with a pair of pens in a pocket protector, and a fancy watch on his right wrist. *Gotta be left-handed.*

Emmanuelle introduced herself and shook hands as firmly as Ben had.

"Well, the name sounds French," Jake said, "but I'm not detecting any accent." He felt his brows furrow. She had striking raven hair, no lipstick, small diamond earrings, a canary yellow blouse, and clearly, a no-nonsense manner.

Emmanuelle explained that she had named Margaux after a favorite aunt on the French side of the family. "I've visited various parts of the clan in Europe, but never lived there for any length of time—except for six months in Heidelberg when I was in school, so my German is competent. Sadly, my French is not comparable, as my familial partisans of that language keep sweetly reminding me."

Jake, now experienced in carrying on conversations at an inn, asked what she had studied.

Emmanuelle replied, "Physics, undergraduate at MIT, a master's at Columbia."

"And what drew you to physics?" *This lady has some impressive credentials.*

By this time, Ben, after a bit of spirited resistance, had gotten Margaux ensconced in the high chair.

"Mama, mama," she said, holding both arms up to her mother for help. Emmanuelle calmed her with one hand on her forearm and another on the opposite cheek.

Emmanuelle replied, "I'm working toward a degree at the Institute for Nuclear Studies at George Washington—"

Ben interrupted, pointing out the oatmeal, which was almost boiling over.

"Abbey told us last night both of you were at GW university as well," Emmanuelle said. "I'm also doing some work for the federal government."

Jake, distracted by the oatmeal, turned off the induction stovetop, moved the pan to a cooler spot, lifted the lid for a couple of seconds, then set it back down. Still distracted, he said, "But . . . your area is in physics. What's your area of concentration?" He realized he needed to get some food to the baby first, so he asked if dry Cheerios would be acceptable. Margaux went after them as soon as he dropped a few pieces on her tray. Emmanuelle got busy getting a bib on the little imp, while Jake retrieved the couple's own applesauce from the fridge and Ben sat next to the high chair to ladle it out to his daughter with a small spoon.

Emmanuelle resumed. "You're probably familiar with the Institute there, which deals with nuclear science, technology, and security, all of which I think are important." She sat in the chair opposite Ben, who continued tending to the toddler.

Jake spoke up. "It's a pretty big school, so perhaps no surprise neither of us ever crossed paths with you on campus. A small band of us started an energy group last year, and we have a seminar scheduled in November with Enoch Apfelbaum. Do you know him, by any chance?"

"Yes, absolutely," she replied. "Now, tell me about this energy group."

Jake served up the salmon frittata, plus a bowl of oatmeal to Ben per request, and then from the fridge brought back some hot sauce for the eggs and papaya juice, which he put in a small pitcher on the table. Emmanuelle suggested to Ben he offer a little of the juice to their daughter, who predictably produced a comical face and dribbled juice down her chin to the bib.

Jake laughed. "Well, first of all, the energy group morphed into a joint effort between three departments—geology, physics, and engineering. But, hey, now we have a couple of classes in the curriculum, even several public lectures. As the newest kid on the block in the engineering department, one of my tasks has been to take on an administrative role for PRICE, which stands for Program for Re-Integration of Climate and Energy. Not my favorite acronym," he said with a grimace, "but at least it's pronounceable.

"I've been interested in energy issues since I was a kid . . . not sure why. I grew up in Minneapolis, only three blocks from some railroad tracks, so I've always loved trains. Maybe that was part of it. Had a big model railroad set up in my basement, HO scale. Trains, you may know, are about six times more energy efficient in transport per ton-mile than trucks are." *So fun talking with other nerds.* He decided to serve himself some oatmeal and a glass of the papaya.

Thinking back, Jake realized he'd failed to pick up on Emmanuelle's reference to nuclear security. Turning to her, he asked, "What sort of work do you do with the feds? Is it with the Department of Energy, by any chance?"

Emmanuelle paused almost imperceptibly. "As a matter of fact, yes."

"But not with the Department of Homeland Security, though?"

"No."

"I would then guess something like NEST, the Nuclear Emergency Search Team, or some other allied bureaucracy," he said as he took a bite of oatmeal.

Emmanuelle clarified one point. "The S stands for 'support' now, not 'search.' Also, NEST is really an umbrella group encompassing all the nuclear emergency response capabilities of the DOE's National Nuclear Security Administration. It involves a lot of specialties, like spectroscopy, radiography, and atmospheric modeling. Some folks think of them as a convocation of nerds,[4] but they've got former military personnel, plus specialists in maintenance, logistics, and communications." She stopped for a moment. "Not . . . like what you see in the movies," she said, with a subtle wrinkling of her eyebrows.

"Sounds like you have more of a problem with awkward acronyms than we do," Jake said, scratching an elbow.

Ben laughed and Emmanuelle grinned and shook her head.

Jake continued. "A lot of my work is on the North American electric grids, and there are security issues there, for sure. Cybersecurity for example, which is probably in your bailiwick, Ben, right?"

"Cyber hardening comes first," Ben replied. "Defensive stuff, not just in the grid, but corporations, the military, that sort of thing. Obviously, cybersecurity is paramount in lots of sectors, like communications, health care, water, agriculture . . . let's see . . . manufacturing, chemicals, and utilities. But I don't work for the government directly, only in consulting."

Jake smiled. "Not the usual sort of conversations you hear in an inn, I gotta say. But then, we get a lot of folks from DC who come out here just to get a break from their work, as in your case, I'm assuming. In fact, we meet an awful lot of out-of-the-ordinary individuals."

"So how is it that a couple of young people like you are running an inn?" Ben asked as he spooned some hot sauce on his frittata.

Jake held up his finger as he swallowed, then replied, "I guess Abbey didn't have time to tell you about our situation. Her parents Anne and Colby established the Inn a dozen years ago, but periodically work overseas in the foreign service. They now have almost another year to go in Abu Dhabi. Usually, they hire a professional innkeeper or inn-sitter, but this time the stars fell into alignment and they offered us the opportunity," Jake said. "I was about to finish up my doctorate, of course. Abbey was in her second year of a fellowship in infectious disease—also at GW—so between the two of us, they figured we had enough time to keep the place up and running."

Emmauelle laughed and shook her head. "You're both really busy and Abbey told us she's due in March. You must be as excited as we were two years ago," she said, as her face softened into a smile.

Jake nodded. "I've gotta ask both of you an odd question. How has the whole experience of having a kid been different somehow than what you thought it would be?"

"Our whole survival," Ben said, "depends on our nanny, Karina, who is wonderful. However, this week she had a family emergency back in Warsaw, a grandmother dying unexpectedly, but we decided to keep this reservation anyway. Margaux almost always sleeps through the night, though she had difficulty last night adjusting to the novel setting. I hope she didn't bother you two."

Jake shook his head, reassuring Ben that she was no bother at all.

"We were lucky to find a place that accepts kids," Emmanuelle said. "Many do not, you know."

Jake scratched his elbow again. "Abbey's parents have always had that policy, which is good, I think."

As they chatted, the room was filled with the babbling of Margaux and her father's intermittent responses. It wasn't often that Jake was able to talk to others in his field over breakfast, and his curiosity was piqued. "I'm curious. Can we go back to this issue of nuclear security? You know, both physical and cybersecurity."

Jake, as usual, pursued his point like a bulldog. "My friend Brian told me a nuclear 'reactor pressure vessel'—surrounding all the uranium fuel rods—is constructed with hardened specialty steel about twelve centimeters thick, almost five inches."

Ben asked what that looked like, but it was Emmanuelle who explained that he would never, ever see one, because it was nested inside a larger "containment" building, often in the shape of a dome, or sometimes a cube. This larger structure, she explained, protects the pressure vessel and all the radioactivity safely contained within it.

Jake took over from there. "The much taller building located close to the containment building, kind of shaped like an hourglass—which many folks confuse with the reactor itself—is just the cooling tower tasked with dumping waste heat into the atmosphere in the form of steam, nothing radioactive."

Jake took another sip of his juice. "As I understand it, the whole containment building is air-sealed and built to survive the impact of a plane crash. Obviously there's a security zone around the whole structure with three successive rows of fencing, plus a restricted airspace above it."

He turned to look at Ben. "Finally, cybersecurity. Again Ben, this is back to your area."

But it was Emmanuelle who answered. "You're preaching to the converted here. One of the basic principles is an 'air gap,' which means a separate computer network within the whole reactor complex without any connection to the outside world—"

Jake interrupted her, knowing as he spoke that this was one of his bad habits. "But you can't be completely disconnected. You have to have phone and internet communications with the electric grid, and with outside emergency services like fire and police, right?"

"Granted," she said, "but all those electronic communications are separated from the control systems for the reactor, heat exchange components, backup diesel generators—all the working parts of the nuclear plant."

Jake was nodding, thinking he should be taking notes.

Ben wiped Margaux's face. "Every time you hear about a reactor emergency, seems like one of the four diesel generators fails to start up."

Emmanuelle frowned. "That's why there are four of them for each unit and they undergo regular maintenance and testing. Even one would suffice to protect the reactor as it's shut down into safe mode. That's not the most critical link. The reactor containment building itself is hardened to kinetic attack like a crashing plane or a drone attack, but most of the heat exchange apparatus lies outside that protection."

Jake knew Ben wouldn't necessarily understand that the water going through the reactor ran in a continuous loop and traded off the heat to a separate plumbing loop outside the containment building. He explained that the heat of the second loop was used to produce steam, which rotated turbines, which generated electricity. Once you got to steam, then nuclear reactors, coal-burners, and natural gas generators all worked the same way.

And all thermal technologies wasted about two-thirds of their energy as discarded heat.

Emmanuelle said, "There has to be absolutely continuous cooling of the reactor core, because even if all the control rods have been inserted to stop the atom-splitting process, the nuclear fissioning. Still, the heat from residual radiation is enough to raise the core temperature far above the operating limits and can lead to melting of the fuel assemblies."

"What're you saying?" Jake said.

"Simply that either a kinetic attack, such as a missile, or a cyberattack could target the grid connections and diesel array or the heat exchangers and steam generators. In other words, if you cut off inside and outside power—which is called a 'station blackout'—or, if you have a loss of coolant accident, then . . . well, you're up the creek without a paddle."

At this point, Margaux voiced her desire to get out of the high chair. She poorly tolerated one last face-wipe and was off and running, Ben chasing after her.

"I always wondered," Jake said to Emmanuelle, "if it was worse to be up the creek or down the creek without that paddle. If you were up, then at least you could figure out a way to get down, but not the other way around."

He picked up his bowl and started piling everything in the sink. "So, where are you folks off to today?"

"We noticed there's a park for Margaux right across the street. I think we'll start there, perhaps grab lunch later. Then she'll need a nap, so we'll come back. After that, perhaps you could suggest a restaurant for dinner, unless we've figured out something for ourselves."

"Nice having a couple of unscheduled days."

"Tell me about it," she said as she stood to get ready for their child's adventure.

FIVE

Resuscitation

The first breath of life is [an] inhalation. The last breath of life is [an] exhalation.
—*CIA*, World Factbook

Death is fugitive; even when you're watching for it, the actual instant somehow slips between your fingers.
—*Jonathan Stroud*, The Whispering Skull

In rapid succession, the receiving ER team at GW took the report from Kristy, lead paramedic, drew arterial blood samples, orally intubated the lady in the gold dress, and started a second large-bore IV access point, all while simultaneously continuing resuscitation efforts. In spite of an initial two liters of lactated Ringer's—plus fresh frozen plasma and platelets to replace coagulation factors and six units of universal donor, type O, Rh-negative blood—attending physician Ned Lasker was ready to halt the endeavor forty minutes in. Lasker, bewhiskered and pudgy, asked the usual, teaching-hospital question, "Anybody have any more ideas?" He looked around at the circle of nurses, residents, a couple of med students, and one respiratory therapist, but saw only heads shaking. "All right. I think we're done here. Now let's see if we can figure out what happened."

ER trauma rooms were typically covered in blood after such an exercise, but this scene seemed bloodier than usual. All of the patient's puncture sites, including the blood draws, had continued seeping blood throughout their efforts. Her oropharynx had a pool of blood, and the inside of her endotracheal tube was laced with rivulets of crimson. The residents, nurses, and respiratory therapist all took great care in stripping off their blood-smeared gloves and gowns, discarding them into the large medical waste container. Only after scrubbing and drying their hands thoroughly did they drop their masks.

"How customary is it to have that amount of bleeding?" asked one of the med students, Hannah Whitlock by her name tag. She pointed at the breathing tube projecting out of the patient's mouth. Hannah was relaxed but intent and soft-spoken.

The other attending physician, Peter McBride—short, intense, and sporting a dark ponytail—responded as he stripped off his gloves and gown. "It wasn't a traumatic intubation—you saw that. The blood in the endotracheal tube is from her underlying bleeding disorder. And we have laboratory data confirming her underlying coagulopathy. Tell me, Hannah, what do you think caused this train wreck?" Both McBride and Lasker knew the importance of pimping the med students.

"With her fever and bleeding and scarring from injection drug use," Hannah said, "I'd think first of sepsis with secondary DIC."

Doctors Lasker and McBride nodded, but it was McBride who answered. "Never narrow your differential diagnosis too quickly. But you're right, disseminated intravascular coagulation is a good possibility. Whatever the problem, the protein coagulation factors and the cellular platelets get used up faster than they can be replaced. Where do extra platelets get mobilized from?" he asked, looking at med student Melanie Drexler this time.

Melanie tucked a lock of blonde hair streaked with green behind one ear and responded quickly. "Half of them from the spleen when under stress, also bone marrow, and along endothelial cells lining the capillaries. By the way, I'd add infection of a heart valve to the differential diagnosis, as she really is a classic presentation for a 'shooter with a fever.'"

"In that case, with endocarditis secondary to drug use, which of the four heart valves would most likely be involved, and why?" asked McBride as he continued to shed his own gear.

Melanie answered, "Tricuspid valve, because that's the first right-sided valve exposed to blood returning to the heart from forearm veins in these drug-abusing patients." She paused. "Is there anything that would argue against Ebola virus in a case like this?"

"Good thought, except there's not been any report of Ebola in North America for a couple of years now," Lasker said. He also knew that most out-of-hospital resuscitations failed.

The team performed a quick review of meds and measures used since the sad lady with the besmeared dress reached the ER, but could not come up with alternative, life-saving treatments. Even cracking her chest to perform direct heart massage would likely have done little to change the outcome.

"Dr. McBride and I have to keep moving," Lasker announced, "just as you need to get back to your assigned patients. I'd suggest you two attend the postmortem if you have a chance, since you're in the ER this month. Notice we leave all the established lines in place, clamped, as well as the endotracheal tube, in preparation for transferring the patient to the morgue for an autopsy." He turned when he reached the door. "Another thought . . . if this turns out to be sepsis as suspected, you might want to run the case by the infectious disease team. There's a new ID specialist by the name of Abbey London who is supposed to be very sharp. You could bring up your idea of Ebola with her. More discussion later."

The two students reviewed the patient's lab information before heading off. Her blood gases showed a low pH of 6.95 compared to a normal 7.4, indicating a slim chance of survival. Her chest X-ray showed multiple fluffy infiltrates, more in the left lung. Her hemoglobin was an anemic 7.2, but once equilibrated with all the fluid they'd poured into her, it probably would have been worse. The platelets involved in blood clotting were considered acute phase reactants, and ordinarily increased to combat severe injury or illness. But her count was abysmally low at 23,000.

Melanie reminded Hannah what Winston Churchill had called the Russians in 1939: "A riddle, wrapped in a mystery, inside an enigma." Somehow, the phrase fit.

Attending the postmortem would be a superb idea.

SIX

Quantum Weirdness

*No matter how hard we try, no matter how much training, how many
safety devices, planning, redundancies, buffers, alarms, bells and whistles we
build into our systems, those that are complexly interactive will find an
occasion where the unexpected interaction of two or more failures defeats
the training, the planning, and the design of safety devices.*
—*Charles Perrow,* Normal Accidents: Living with High-Risk Technologies

*With a fourth generation of nuclear power, you can have a technology that
will burn more than 99 percent of the energy in the fuel. It would mean that
you don't need to mine uranium for the next thousand years.*
—*James Hansen*

*For 50 years, nuclear power stations have produced three products which only a luna-
tic could want: bomb-explosive plutonium, lethal radioactive waste, and electricity so
dear it has to be heavily subsidized. They leave to future generations the task, and
most of the cost, of making safe sites that have been polluted half-way to eternity*
—*James Buchan*

Rachel was running a bit behind schedule for this evening's lecture,
but still managed to find one of the open seats on the third level
up in Lehman Auditorium, located on the first floor of the Science
and Engineering building of GW. She draped her coat over the back of
the chair as she sat and proceeded to organize the gear from her backpack,
boot up her laptop, and place her water bottle on the floor. Abbey came
in next, interrupting Jake for just a moment. He pointed up toward Rachel,
situated just shy of the top row, so that Abbey could join her. He was too
distracted to do much more than smile. *Here I go again. I need to calm myself
down.*

Jake was excited to be introducing the speakers, his first official act as a newly minted associate professor of engineering. And since his department had decided to open some presentations to the public, the crowd was larger and a bit more boisterous.

The next two people in the door were newcomers. The older one, he guessed, was probably in his late thirties, with a moderate, trimmed ruddy beard. The other appeared younger, fit, and clean-shaven.

Rachel pointed at the younger man and whispered, "Tall, dark, and handsome," as Abbey sat down.

Jake introduced himself to the two speakers, shook hands, and turned to see an appreciably larger turnout. He stood to one side of the pair and took a big breath. "Evening folks, thanks for being here tonight. I'd like to introduce Professor Enoch Apfelbaum from the physics department, who'll be our main speaker tonight. With him is Noah Williams, a first-year graduate student with a special interest in lasers and optics. I'm confident we'll welcome their participation and perspectives here tonight. And, with that—let me expand everyone's vocabulary by saying I will now officially 'convoke' this meeting."

A few dismissive guffaws followed as Jake sought a seat in the front row, glancing up and smiling at Abbey before sitting down. He took note of a guy in the very back row wearing sunglasses, which seemed curious for an evening class indoors. *Takes all kinds, I guess.*

Professor Apfelbaum spoke first. "First of all, I know my last name is a mouthful. Call me Enoch. Professor Higgenbotham has indicated that you'd like to keep these discussions informal, and I'm in full accord. Feel free to speak up at any time if you have a question or comment."

"We're all aware that nuclear power creates a fair degree of ambivalence, even among environmentalists. The other energy sectors slot into the categories of the good, the bad, and the ugly pretty unequivocally, but nuclear power has elicited strong controversy, even in my own department."

Jake opened his notebook to take notes. He'd been looking forward to this lecture for weeks. Enoch picked up a black marker and wrote in strong script on the whiteboard:

Nuclear Proponents

"On one side are advocates who argue nuclear power is critical for transitioning to a carbon-neutral future. That it's a key pillar of baseload power in the electric grid. That it has the advantage of being homegrown. That the risks of radiation-release accidents are overblown. That our only consequential reactor meltdown at Three Mile Island resulted in no discernible health effects. And that our country has more operating commercial nuclear plants than any other nation and should not squander this lead and expertise."

Back at the whiteboard, Enoch added underneath:

Nuclear Opponents

"On the other side are those detractors who just as vociferously refute these same points. They argue that trying to increase the size of our nuclear fleet will impede the development of renewables in what is, after all, a zero-sum funding game from the governmental and financial sector. That attempting to establish an all-of-the-above energy strategy, as Obama did, was a strategic mistake." He picked up the remote from the lectern and turned on the projector.

"That it's not solar and wind which are unfairly subsidized by the federal government, but rather that nuclear power has been benefiting from a grander set of subsidies for almost seventy years, in what should now be considered a mature technology. That nuclear is baseload power in the grid only because it's too brittle to easily modulate its power output." Enoch paced back and forth across the front of the room, gathering momentum as he ticked off his points.

"That there were never adequate epidemiological studies after Three Mile Island to ascertain harmful health effects." Enoch slowed his pace.

"That the number of reactors is, if not plunging, then steadily being eroded away by economic competition, not public opposition. That the risks of radionuclide release occur at every stage of the nuclear fuel cycle, from mining and milling on the front end all the way to long-term radwaste storage on the back end. That the proposed permanent repository at Yucca Mountain in Nevada has more geologic faulting and potential water infiltration than initially appreciated and is thus unsuitable for the jettisoning of nuclear waste."

Jake made a fist in his lap. *Absolutely agree.*

Enoch stopped and turned in the center of the room to face the audience. "Finally, opponents argue that nuclear power plants are vulnerable to cyber intrusions and kinetic terrorist strikes, and in wartime may well be considered prime targets for conventional warfare; hence, it's strategically unwise to construct them."

Jake's first thought was of Ukraine. His second was that reactors are more hindrance than help in structuring a twenty-first-century electrical grid.

Enoch began writing again:

Nuclear Plant Accidents

"A number of serious incidents involving the nuclear industry have occurred, as you can see in this first slide," Enoch said. "Most proponents mention only the big three in Japan, the US, and Ukraine—back when it was part of the USSR. But the list is considerably more extensive. In fact, almost a hundred reportable civilian and military nuclear power plant accidents took place worldwide, from 1952 to 2009. By definition, these misadventures either resulted in mortality or more than fifty thousand US dollars of property damage." Enoch pressed the remote and the first slide lit up next to the whiteboard.

Major Nuclear Reactor Catastrophes

Chernobyl	Lucens[5]
Three Mile Island	Fukushima Daiichi
Rocky Flats	Tokaimura[6]
Kyshtym	Marcoule
Browns Ferry	Windscale
Idaho Falls	Bohunice
Mihama	Church Rock

"This compilation is limited to nuclear power plants, excluding weapons incidents. The adverse outcomes have included commonplace regular tritium releases, mine-tailing radioisotope releases, nuclear waste dispersals, trafficking in plutonium and other radionuclides, thefts of special nuclear material, and many more."

Jake tapped his notepad with his pen. *Damn, I don't recognize half these names.*

His pal Brian spoke up. "Professor, could you mention something about some of the more obscure names, like Idaho Falls?"

Enoch turned to face him. "We realistically don't have time to discuss the whole list, but I can tell you the story of Idaho Falls, which is historically obscure. This was a small prototype nuclear plant in Idaho, called SL-1, in which the US experienced its first fatal nuclear accident. By 1961, twenty-four experimental reactors of various types had been constructed in the vast National Reactor Testing Station in the remote high desert of eastern Idaho. SL-1 had the ominous combination of a design flaw and a troubled technician. A power surge wrecked the building and killed three men, widely contaminating the area with radioactivity. An investigation revealed two of the men may have been involved in a romantic triangle with one of their wives, that this was a suicidal sabotage, but it was never proven." Turning back to the board, Enoch erased all prior headings and wrote:

Evidence of Nuclear Decline

"A number of lines of evidence support the argument of a flagging world nuclear industry. A Paris-based consultancy puts out an annual publication called *The World Nuclear Industry Status Report*,[7] an invaluable and highly respected source of information, as you can see in this next slide. Mycle Schneider is one of the prime movers in this organization." Enoch clicked the remote and pointed to the new slide.

The World Nuclear Industry Status Report 2022

Trend indicators in report suggest world may have reached a number of historic maxima:

- 1976 — peak construction starts
- 1979 — maximum number reactors under construction
- 1996 — maximum share nuclear power in electricity sector
- 2002 — peak number reactors in service
- 2006 — pinnacle of nuclear power generation
- 2019 — decline to one less reactor from 1989
- 2022 — 412 reactors in operation, 25 in long-term outage, 55 under construction

---

"In contrast, in 2021 alone, worldwide renewable energy capacity surged by 290 gigawatts or GW, each gigawatt equivalent to the rating of a typical commercial nuclear generating plant."

Jake thought to himself, *Some 55 nuclear GW in construction—each taking a decade or more to complete—damn, renewables added 290 GW in just one calendar year. Faster, easier, cheaper.*

"These and other controversies make for a fascinating examination of facts and strongly held tenets, but we need to back up a bit and look at the basic science. For this purpose, I've recruited Noah Williams, one of our graduate students, to kick off our discussion. Noah." Enoch held out an arm, welcoming the young man to the front.

Noah casually placed his notes on the lectern. He'd just finished writing the four fundamental forces in the universe on the sideboard:

Gravity

Electromagnetism

Strong Force

Weak Force

"The universe is constructed of all of myriad types of particles interacting with one another through four fundamental forces: gravity, electromagnetism, the strong force that binds an atomic nucleus together, and the weak force that produces a kind of nuclear disintegration called 'beta decay.' Physicists remain uncertain over ever achieving a grand reconciliation of these fundamental four."

Jake didn't write these basic facts down. Instead, he concentrated on the speaker.

"Gravity has not yet been incorporated into a theory of everything. And the latter two forces are limited to operating at the interior of an atom's nucleus. An atomic nucleus is defined by its number of protons featuring a positive charge. A strong force is required to cluster them all together and overcome their overwhelming tendency to repel each other." He demonstrated by pushing his fingertips together.

"The number of protons succinctly defines the identity of any element. A carbon atom, for example, has six and only six protons, forever and ever."

Rachel leaned over to whisper something about the good-looking lecturer to Abbey, who answered her quietly, smiling.

"The other particle in the nucleus is the neutron, which has no charge, but is absolutely required to hold the nucleus together. The number of neutrons is variable in any given element, and therefore, each additional neutron creates a different so-called 'isotope' of that substance. As our working model, let's begin with carbon, which has anywhere from two to sixteen neutrons, with six being by far the most common. Combining the six protons with a variable cohort of neutrons leads to isotopes from carbon-8 up to carbon-22, but most are vanishingly rare. The two common, stable, naturally occurring isotopes are carbon-12 and carbon-13. But carbon is always carbon, element number six on that periodic table of the elements posted on the wall of your high school science class."

The crowd gave a short laugh.

"Conversely, the rare and unstable isotope carbon-14 is notable for possessing a fickle nucleus and is said to be 'radioactive' and subject to spontaneous 'decay,' that is, the nucleus tends to break down at some unpredictable moment, releasing heat and beta radiation. Hence, its identification as a 'radioisotope,' short for radioactive isotope. About 118 elements have been discovered or created over the last two centuries, but the cumulative number of isotopes is estimated in the neighborhood of five thousand."

Jake shook his head at all of this new information and began taking notes again.

Noah posted up a curiosity:

Really Empty Space

Where is he going with this? Jake wondered.

Noah continued. "A nucleus is vanishingly small compared to the size of an atom. If we were to consider a nucleus as a baseball placed in the middle of a stadium, not at the pitcher's mound, but out past second base in shallow center field, then the electrons would be running around high in the bleachers, even in the largest American stadium—home of the Los Angeles Dodgers—which holds fifty-six thousand spectators. Growing up, I watched a lot of games there. But the nucleus would not really scale to the size of the baseball, more like a bit of chewing tobacco or the husk of a sunflower seed stuck to the leather—really, that small." He gave the audience a moment to think about this.

"The nucleus is so small that an atom is about 99.999 percent empty space, which is why you and I can look right through the mixture of gases in our atmosphere—absent fog or smog or vog in volcanic areas like Hawaii—a clear liquid like water, albeit with some issues with refraction at the surface, and even a transparent solid like window glass." Noah halted to post another point:

First Nuclear Reactor

"Let me shift gears—reverse gear in this case—and throw out a question: When and where was the first nuclear reactor in the world?"

Hands shot up. Noah pointed to a fellow in back, who said, "The reactor under the stadium of the University of Chicago, part of the Manhattan Project, I think, in . . . 1942."

Noah just smiled. "You're correct about the year the Fermi reactor first went critical, but you're off . . . by several billion years . . . as far as the first reactor's concerned."

Rachel snorted at the jest, then whispered an apology to Abbey.

"Let me explain," Noah said, noting the sounds of disbelief in the audience. "The other key element we need to discuss today is uranium, element ninety-two on the periodic table, much further down than carbon. And the key uranium isotope is U-235, because this one can be split apart by a single stray, disruptive neutron arriving. This is called 'nuclear fission,' which takes place in a nuclear reactor. Physicists borrowed the word 'fission' from biology, which describes cell division. So physicists say U-235 is a *fissile* isotope. Got it?"

Jake whispered into the air. "The famous splitting of the atom. Releasing all that energy plus a couple more neutrons, yeah, we get it."

Noah asked, "And why is U-235 fissile? It has ninety-two protons defining its identity as uranium, but way more neutrons trying to keep the whole unstable nucleus together. And all it takes is one extra party-crashing neutron to upset this whole apple cart. The proportion of U-235 in natural deposits nowadays is only 0.72 percent, but several billion years ago, that figure was five times higher at 3.68 percent. In what is now Gabon,[8] formerly a French colonial possession in western Africa, algal mats are hypothesized to have concentrated natural uranium ore eroding out into a river system."

Jake half-remembered reading an article about spontaneous fission of uranium in some journal.

"Enough uranium-235 was concentrated in several locations that fission, or atom-splitting, initiated spontaneously. This occurred for two simple reasons. First, the requirement was met that the fissile isotope be in the range of three to five percent. Second, the water served as a moderator,[9] slowing down the 'fast' neutrons emitted in the splitting process. Usually only the resultant 'slow' neutrons are capable of fusing with the nucleus of the next intact U-235 atom. These collisions and fissions amplified exponentially into a chain reaction. Two to three more neutrons released from each split of a U-235 atom, along with huge amounts of gamma radiation and thermal, or kinetic, energy."

"Noah, why don't you explain exactly where those loose neutrons come from?" prompted Professor Apfelbaum.

Nodding, Noah responded, "Like a fast neutron, I probably started out too hastily and got ahead of myself. When an appropriate neutron joins the nucleus of a U-235 atom, all of the suddenly unstable protons and neutrons rearrange themselves, like a phone booth stuffed with one too many college students. Something has to give."

Back when there were phone booths, Jake thought, impressed with Noah's use of humor.

"If the last student arrives too quickly, he may just bounce out of the booth. But if he insinuates himself all the way in more carefully, everybody realizes they can't breathe and they all pile out at once, with two or three of them running away like those stray neutrons. Crude metaphor, but then . . . college students are often pretty crude."

Enoch appeared mildly puzzled.

I get it, great teaching point, Jake thought to himself.

"Enough of a chain reaction occurred that an estimated one hundred kilowatts of heat were produced, equivalent to that emitted from about a thousand large, incandescent bulbs, but only about one-ten-thousandth the power of a modern nuclear plant. Sufficient, however, to boil off the water, stalling or halting the whole process. It's now felt that this unique watershed in Gabon was remarkable for hosting sixteen or seventeen different natural, recurring nuclear geysers over a million years or so."

Noah paused to glance at his notes.

Jake sensed he was trying to figure out the art of transition.

"Enoch has asked me to discuss one last topic: six special capabilities of neutrons. The first three are neutron moderation, reflection, and transmission. The other three are neutron capture: with fission, without fission, or with transmutation. We aren't going to discuss these in great detail, you'll be glad to hear. Moderation is first. Either water or graphite may slow down a fast neutron, thereby creating the slow neutrons needed for fission. Remember that campus phone booth. Reflection is second. The nucleus of beryllium—element number four in the periodic table—causes neutrons to bounce smartly back, and as you might imagine, this is a great element for keeping neutrons corralled inside a controlled space."

Like jailed miscreants awaiting trial, Jake reflected.

"Transmission is third. The nucleus of zirconium—element number forty in the periodic table—has what physicists call a really tiny 'cross-section,' so it's a good element for facilitating free passage of neutrons between adjacent fuel rods."

Zirconium is about as transparent to neutrons as glass is to photons of light, Jake mused, and liked the comparison so much he wrote it down.

"Capture is fourth. The nucleus of cadmium—element number forty-eight on the periodic table—can absorb or glom onto a neutron, converting itself into a heavier isotope, but with no fission or radioactive decay as a result. This means cadmium can be used to squelch an ongoing fissioning process, typically in the control rods of a reactor."

Jake wrote, "Cadmium puts on the brakes."

"Fission is fifth, already discussed. A fissile isotope absorbs a slow neutron, then successfully splits into a variety of unequal fission fragments. This is what nuclear power is all about.

"Transmutation[10] is sixth and last. Uranium-238 is a 'fertile' isotope, able to capture a neutron and be 'transmuted' into an altogether different element. If uranium-238 gains a neutron, it can convert the neutron into a proton, and in several steps, transition into bomb-useful plutonium-239. Which allows the construction of smaller, lighter nuclear weapons, such as the one used against Nagasaki in World War II. I know this goes against our desire to have consistent nuclear particles, but this is their reality. And now, I'll stop right here, unless there are any questions."

Brian spoke up again. "Sounds like transmutation is an actual embodiment of the philosopher's stone—turning lead into gold was the supreme goal of alchemy, right?"

"Interesting you should say that. Ironic even, that Glenn Seaborg,[11] winner of the Nobel Prize in Chemistry in 1951, reportedly succeeded in transmuting a vanishingly small quantity of either lead, or perhaps bismuth, into gold back in 1980. Of course, forcing a stable element like lead to give up three protons requires a huge effort and energy expenditure. The transmutation is far more expensive than the value of the gold produced. Not something you'd want to try at home, anyway."

Enoch had rejoined him at the lectern and said heartily, "Excellent work, Noah, I can see and appreciate how much you like to teach." And with that, the younger man found a seat next to Jake.

Enoch used the transition time to print a name on the board:

Becquerel

"In 1896, Henri Becquerel discovered the radioactivity of uranium. The conventional but erroneous thinking was that radioactivity required uranium crystals' initial exposure to light, that the phenomenon was simple phosphorescence. He performed a series of experiments seemingly supporting this hypothesis, until one cloudy week in Paris he left the uranium—without exposing it first to sunlight—sitting on a photographic plate wrapped in black paper in a closet. Becquerel, for unexplained reasons, then developed the plate anyway, and was surprised to find an identical outcome, indicating the photographic emulsion was receiving a spontaneous, nonsolar source of energy. Further investigation revealed the emissions were continuous and did not measurably diminish with time."

Enoch bent to write a second name:

Rutherford

"Three years on, in 1899, Ernest Rutherford discovered alpha and beta 'rays,' which ultimately turned out to be alpha and beta particles. This New Zealand-born British physicist is considered to be one of the greatest experimentalists of his age, and he won the Nobel Prize in Chemistry in 1908, in part for his work on radioactivity."

In bold graffiti, Enoch scrawled a third name:

Curie

"Marie Curie, a Polish-French physicist and chemist and former graduate student of Becquerel, together with her physicist husband Pierre, went on to discover two other alpha-emitting elements, namely, polonium and radium. She eventually won two Nobel Prizes for the work, shared with both her husband and former teacher. She was also the one credited for coining the term 'radioactivity.'

"Unfortunately, the medical risks were not appreciated by these early researchers. Marie's husband died in a horse cart accident attributed to his declining health from radiation toxicity. Marie herself died of aplastic anemia years later. And their scientist daughter died from leukemia secondary to polonium poisoning. Marie's papers to this day are kept in lead-lined boxes, and are handled only with stringent precautions."

Enoch took a moment to erase and write:

$$\alpha + \beta + \gamma + Neutron\ Radiation$$

"Clearly, radiation is not a unitary phenomenon. Dozens of odd creatures are in this menagerie. But we'll limit our scrutiny to the alpha, beta, gamma, and neutron categories. First, an alpha particle comprises two protons and two neutrons, identical to a helium nucleus stripped of its pair of electrons. Alpha particles are ejected from radioactive atoms on the order of five percent of the speed of light, or about 9,300 miles a second. An alpha particle has large mass, high speed, and a double positive charge—but paradoxically, low penetrating power. The explanation for this apparent paradox is the alpha particle's strong interaction with and absorption by any substance traversed, whether that substance is paper or skin of even a few dozen centimeters of air. But if a 'hot particle'—a microscopic agglomeration of alpha-emitting atoms—is inhaled or ingested, it can lodge in tissue and wreak havoc, with multiple hits on large molecules, such as RNA and DNA. Even a single—and I emphasize single or solitary—microscopic hot particle can be cancer-inducing, or carcinogenic."

Enoch replaced his prior handwriting with the following:

Beta Radiation

"A beta particle is a horse of a different color and exhibits three sub-types. The emitted particle is either an electron or its antiparticle equivalent, a positron. The third type is an arcane entity called 'beta inverse,' and I think we'll not delve into that here."

Underneath "Beta Radiation," Enoch added:

Gamma Radiation

"Perhaps a non-particle type of radiation might be easier to grasp. Gamma radiation is electromagnetic radiation—or waves of very high frequency—and therefore extremely high energy. But it has no mass and no charge, is highly penetrating, and is often released in conjunction with alpha or beta radiation."

Pivoting, Enoch completed his list at last, adding:

Neutron Radiation

"Critically, the fourth type of scattering in nuclear power is neutron radiation. Sources of free neutrons include nuclear fission and nuclear fusion, as well as other unique circumstances. Neutrons are even more penetrating than gamma rays. A most peculiar characteristic of neutrons is their induction of brand-new, or *de novo*, radioactivity. Alpha and beta and gamma radiation may cause abrupt damage in biologic tissue, but leave nothing radioactive behind. Neutrons, however, may join and destabilize the nuclei of other elements, which then undergo spontaneous radioactive decay. This is quite problematic in fission and fusion reactors, because neutron effects gradually render all of the original structural components 'hot' in a radioactive sense."

Brian interjected, "I'm sorry, Professor, but you're clearly distinguishing between thermally hot and radioactively hot, correct?"

Jake knew why he'd raised this question.

Enoch raised both palms and gestured toward the young man. "Never apologize for asking a question. Yes, I'm saying just that. Thermally hot, as in 'too hot to touch.' Whereas, radioactively hot is usually imperceptible at first. Moreover, our four types of radiation have different penetrating power."

Brian wasn't done. "So is this transmutation, like Noah discussed?"

Enoch shook his head. "Typically, this represents only an isotopic promotion by the addition of a neutron, resulting in an unstable or radioactive atom, such as the element nickel in hardened steel."

Enoch set down the marker and addressed the class. "This might be a good point for a break. When we get back, I'll explain the inverse square law of diminishing dose with distance."

SEVEN

Close Encounters

*Tricks and treachery are the practice of fools that
don't have brains enough to be honest.*
—*Benjamin Franklin*

J ake went outside for a breath of fresh air. When he turned around, he was surprised to see the guy in the sunglasses standing just inside the building's shadow. *How did he sneak up on me so quietly?*

The stranger wore a black baseball cap and a forest-green, long-sleeved shirt, frayed around the cuffs, with a pocket protector empty of pens. A couple days' dark stubble peppered his pallid face. Eyes shifting back and forth, he scratched his cheek, momentarily covering his mouth.

Jake issued a tentative "hello."

"Your name's Jake, right? I came outside because I needed to know what you thought about nukes. Apfelbaum seems to argue against 'em. What about you?"

"I'd say I've been an antinuke guy since college." *Intensely so, always will be*, thought Jake, but he didn't want to start a debate.

"Yeah? Where was that?" He grimaced and continued to avoid looking directly at Jake.

"University of Minnesota in Twin Cities. How about you?"

"Couple a places, not anywhere around here." His eyes were trained on the ground. "But what I wanna know is whether you think nukes have gotta be stopped, okay?"

Jake bit his lower lip for a moment, wondering where this conversation was headed. "Actually, given the current status of our creaky grid, I'm in favor of keeping operating plants licensed as we transition to renewables. Otherwise, pretty much what Enoch laid out. Plus, I know Finland's the only country actually taking steps to build a permanent depository to bury radwaste deep underground. Not to mention the US military has lost track of the location of their warheads on multiple occasions, like the Broken Arrow and Empty Quiver events."

Jake stopped for a second, still puzzling over this guy, but he was on a roll. "Any country even building civilian nuclear plants might as well paint a target on the containment buildings, because they're obvious targets, like what happened in Zaporizhzhia."

Jake stared him down. "Ukraine showed us how the breach of plant with a conventional weapon could trigger a nuclear attack through the widespread release of radiation."

The guy was tapping his foot, head bobbing in agreement.

"Small, modular reactors are unproven and require higher-enriched uranium than conventional reactors—even more of a proliferation risk— and feature hard-to-service heat exchange units that lie inside the reactor pressure vessel itself. And their projected cost, when and if they get produced by the end of this decade, is almost three times that of solar PV, plus storage, or land-based wind." Jake realized he was rambling. "And you? What do you think about all this?" He wasn't sure what to expect from this guy.

There was a pause as the stranger shuffled his feet again. "But don't you think we should do something about nukes, really do something . . . like now?" The man's hands made fists. "Okay, okay, that's all I needed." He turned around and went back inside.

What the hell was that?

Squaring the Circle

A few students and visitors were trickling in as Enoch got under-way. "Okay, let's get started again. For a single point source of radiation, the 'inverse square law' explains diminishing dose with distance. That is to say, if an object or person is twice as far away from the source of radiation, the dose is reduced by one-half times one-half, equal to one-quarter of the dose at a closer distance. This slide demonstrates the distinctive penetrating power of these four radiation types." Enoch pointed to the slide.

Jake found it hard to concentrate, still feeling a bit perplexed about the encounter with the odd character.

Penetrating power of different types of radiation

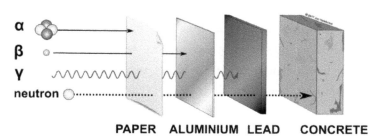

PAPER   ALUMINIUM   LEAD   CONCRETE

Figure 1: Image by Jan Helebrant

"Alpha particles are stopped by paper or intact skin. Beta particles will be halted by a few sheets of aluminum foil, which explains those hats sported by slightly demented people."

Jake recalled an aluminum hat worn as a kid in response to reports of UFO sightings. *More silly than demented.*

"Intense gamma rays are diminished by half for every centimeter of lead, six centimeters of concrete, or eighteen centimeters of water, then half of that half again with repetition of the same barriers. Neutrons, finally, will largely penetrate all these impediments lined up in a row. Think of the infamous 'neutron bomb' that hypothetically would damage only living creatures, leaving infrastructure largely intact."

Enoch circled around the lectern to write, quickly erasing one misspelling after his first attempt at the second word:

Becquerels + Sieverts

"I propose and submit to you two new foreign terms, namely 'becquerels' and 'sieverts.' The number of radioactive particles emitted is counted in becquerels. Biologic effect is measured in sieverts. Henri Becquerel achieved the honorific, defined as one nuclear decay per second, in any arbitrarily sized sample of radioactive material one might be dealing with. Swedish physicist Rolf Maximilian Sievert defined the unit of exposure in the calibration of biologic effects of background, therapeutic, accidental, or military radiation exposure. In this comparison, alpha particles are twenty times more potent and damaging than beta particles."

Jake felt challenged—and suspected others might as well—as he tried to keep up with this torrent of concepts.

With a finger alongside his nose, Enoch innocently asked, "If I offered you a small, lead-lined box containing a contaminated metal thimble exhibiting ten becquerels of beta radioactivity—ten radioactive disintegrations per second—who would be brave enough to open up the box, and pick up the thimble? Show of hands?"

Several were raised, tentatively.

"Okay, then, how many becquerels in a banana, any idea?"

No volunteers, but a room full of curious looks.

Enoch wrote on the board:

Banana Equivalent Dose

"The history of the banana is an ironic one. When the Atomic Energy Commission, or AEC, the forerunner to the Nuclear Regulatory Commission, or NRC, was trying to accustom people to the concept of permissible or background radiation—and persuade them not to fret—they hit upon the idea of using the banana as an informal and educational tool."

Grins appeared among the audience.

Enoch countered with a slow smile of his own. "No. Not what you're thinking. Tools to demonstrate radiation, not condoms. Turns out an average banana exhibits about eighteen becquerels, eighteen decays per second. Yet we eat this fruit all the time without hesitation. I assume Becquerel and Sievert ate bananas too." He smirked at his own jest.

"Bananas contain radiopotassium, potassium-40, so they give off beta particles—of all three classes. But the biological impact? The question you should be asking yourself is, How many sieverts per banana? And the answer is about 0.1 microsieverts. And a microsievert is one-millionth of a sievert."

Blank looks appeared in the audience. Jake broke the silence. "Enoch, how does that compare to background radiation?"

Enoch pointed at him with both index fingers. "An absolutely insightful question. Good. I was actually looking for that. Now, reference the table on this next slide," he said, clicking his remote.

"You can see the amount of background radiation ranges widely from two to fifty millisieverts, or mSv, per year. And a millisievert is one-thousandth of a full sievert." Enoch turned to the class to emphasize his point. "It's important to understand that each individual exposure is *additive*, and under the 'linear no-threshold model' accepted by most scientists, there's no safe lower limit of exposure, and risks comprise at least induced cancers, genetic mutations, and teratogenic effects during pregnancy."

Annual Additive Radiation Exposures

All Figures in Millisieverts [mSv] per year

Worldwide mean background exposure	2
US average natural background radiation	3
Background parts of Iran, India, Europe	50
10,000 bananas or > 27/day [extra ~3,000 kcal]	1
New York–Tokyo flights for airline crew [annual]	9
Astronauts on ISS 6 months	120
Smoking 20 cigarettes/day	9–40
Chest X-ray [2 views]	0.04
Abdominal CT scan	10
NRC max acceptable man-made dose US civilian	1
Residence near nuclear reactor [absent accident]	0.01–0.0001
Living near coal-fired generator	0.0003
Current limit US nuclear workers [annual]	20
Japanese radiation workers before Fukushima	100
Modified limit 3 days after Fukushima	250

Enoch pointed at the slide. "Notice you'd have to eat ten thousand bananas a year, more than twenty-seven a day, to accumulate a single extra millisievert."

Nobody likes bananas that much, thought Jake, shaking his head.

"The AEC arbitrarily set the maximum allowable exposure to civilians at 1 millisievert per year, a standard the NRC has continued, and you can see, unless there's a radiation-release accident, the typical exposure is a tiny fraction of that. Medically, you should never balk at a two-view chest X-ray, especially with the new digital technology. But at least discuss the radiation risk of an abdominal CT scan with your doctor, equivalent to about four years of background radiation exposure."

Jake turned and noted Abbey raise her chin almost imperceptibly, presumably concurring that this was a reasonable public health recommendation.

Enoch continued. "Finally, the US and Japan had set different occupational exposure limits, a five-fold difference. But then, three days after the Fukushima disaster began unfolding, Japan increased the allowable cumulative exposure from 100 up to 250 millisieverts. Again, thousandths of a sievert." He paused.

"There's an overt dichotomy here. Small, incremental doses in the millisievert range on top of natural background radiation create only narrow public health concerns. Doses in orders of higher magnitude, in the sieverts, constitute a medical emergency to be cared for in a tertiary care setting—typically a university medical center."

The listeners appeared rapt but sober. Jake knew Abbey had no pertinent or personal experience, so he assumed she had no serious disagreements. Enoch used this pause to sip his lemonade left-handed as he posted:

Radioactive Decay

"Let's move on to spontaneous, radioactive decay and its rate of decline measured in half-lives. Radioactive 'mother' atoms have decay product 'daughters.' If you're looking at a single atom, physics has no way to predict the timing of its decay. But if you have a larger sample, you can predict what fraction will have undergone decay after a half-life interval, which remains at a constant rate for that particular isotope.

"After ten half-lives, the residual radioactivity will have been reduced to less than one-thousandth of the starting quantity. After twenty half-lives, the remaining radioactivity will have been reduced to less than one-millionth of the initial level. Thus, as a rule of thumb, ten to twenty half-lives will reduce the risk to a presumably acceptable level.

"These invariant, radioisotope half-lives range from extremely long to extremely short. Let's look at a couple of examples."

Enoch reached high on the board to start composing a list.

Cesium-137

"Now, cesium-137, which emits both beta and gamma radiation, tends to concentrate in muscle tissue and has a half-life of about thirty years. So how long an interval is ten half-lives?"

Rachel called out, "About three hundred years, and twenty half-lives would be six hundred."

"Good," said Enoch, then wrote again:

Strontium-90

"What about strontium-90, found so infamously in deciduous teeth shed by American children as a consequence of above-ground nuclear testing in the 1950s, with a half-life of nearly twenty-nine years?"

Jake stated confidently, "About two hundred and ninety years and five hundred and eighty."

"Good at mental math, I see," Enoch acknowledged. Another scribble:

Iodine-131

"Last one now: iodine-131, which concentrates in the thyroid gland, with a half-life approximating eight days."

Brian quickly answered, "Eighty days and one hundred and sixty days."

"Excellent," Enoch said. "Now, here is a challenging philosophical sort of question. Which is the more dangerous radioisotope? Short half-life? Or long half-life?"

No immediate response. Even Jake felt flummoxed.

"Shorter half-life radioisotopes are more intensely radioactive and kinetically—or thermally—hotter, while longer half-life radioisotopes are less radioactive and thermally hot, but are dangerous for eons. So they're dangerous . . . in different ways." He scrubbed his list and marked up:

Nuclear Fission + Chain Reaction

"Finally, we're ready to discuss nuclear fission—the splitting of atoms—which is not decay but rather a chain reaction. The result is not decay daughters but rather much smaller atoms called 'fission fragments,' coupled with huge amounts of gamma radiation and a prodigious amount of heat. In a reactor or a nuclear bomb, imagine a single first-order slow neutron with a miniscule amount of energy on the order of a single electron volt. Don't worry about what an electron volt is, but keep this in mind for what comes next.

"It just so happens that this slow neutron randomly collides and merges with the nucleus of an atom of uranium-235. Instantaneously, the nucleus splits into two asymmetric fragments with one possible outcome being krypton-92 and barium-141, plus two or three free neutrons and a colossal amount of energy. Alternative possible fission fragments include cesium-137, strontium-90, and iodine-131. Sound familiar?"

People were bobbing their heads in agreement.

"Of course they should. And how much energy is released?" Enoch's voice slowed, even deeper in pitch. "Why don't we calculate that? The slow neutron that sets this all off arrives with the energy of a single electron volt. So do the expelled second-order neutrons—once slowed down—that scatter and split other U-235 atoms. Then, third-order neutrons spread out like rugby players bursting out of a scrum, and so on.

"Now, appreciate that the energy released by fission is calculated," Enoch said, his voice rising, "by that most famous equation in all of physics—Einstein's 'energy equals mass times the speed of light squared.' While the number of the protons in the uranium nucleus is identical to the combined number in the fission fragments, the nuclear mass before fission is larger, which is attributed to something called its 'nuclear binding energy.' That binding energy mass is not lost, but instead converted into different types of energy via Einstein's equation, with that quantity multiplied by a huge number twice."

Enoch paused theatrically. "The speed of light times the speed of light. In fact, the total energy ramps up to 215 million electron volts." He held his arms out wide for emphasis.

"So, inordinate amounts of gamma and neutron radiation and kinetic, or thermal, energy are released. In a reactor, expelled neutrons hit other uranium-235 atoms, and the reaction may continue and be carefully monitored and controlled for the production of steam to run turbogenerators for the creation of electricity."

"That's where I come in," Jake whispered to himself.

"Or, sadly, a nuclear device may be detonated as an uncontrolled chain reaction."

Enoch wrote a famous name on the board:

Manhattan Project

"In 1942, that first man-made reactor in Chicago used graphite as the moderator; used cadmium, indium, and silver in control rods; lacked a radiation shield containing beryllium; and utilized no cooling system except ambient air. The leader of this project was the Italian-American physicist Enrico Fermi. The sole, rudimentary safety system against a runaway reaction was an additional backup control rod that could be dropped down from a safe distance into the center of the pile using an ax to cut a rope run over several pulleys. This system was to be used in response to the yelled command 'scram' in the event of runaway fission or other catastrophe. 'Scram,' in other words, meant 'run for your life.'

"The myth grew up later that 'scram' stood for 'Safety Control Rod Ax Man,' but this was purely a fabrication. Regardless, to this day every reactor contains an installed scram control, albeit with neither rope nor ax.

"Eisenhower's Atoms for Peace program after World War II was perhaps designed in part for atonement. Our first commercial reactor was the sixty-megawatt Shippingport reactor in Pennsylvania in 1958, though historically it was preceded by the first small Soviet plant of five megawatts, which went into operation in 1954. Commercial plants nowadays tend to be rated at about one thousand megawatts, which is equal to one gigawatt."

Nuclear Fuel Cycle

"The long, complex nuclear fuel cycle may be split into three parts, with a front end, the reactor in the middle, and a back end. It all starts on the front end with the steps of uranium prospecting, mining, and milling. The work product is called 'yellowcake,' a granular bright yellow material, hence the name. The yellowcake is converted to gaseous uranium hexafluoride by the addition of six fluoride atoms to each uranium atom. This allows concentration of the desired fissile uranium-235 to a useful three to five percent for a power reactor, twelve to twenty percent for a research reactor, or finally, weapons-grade material of at least eighty-five percent."

"The enriched material is converted back to uranium-oxide powder and formed into short cylinders about an inch in length. These pellets are stacked into twelve-foot-long fuel rods in tubes of zirconium alloy cladding. Then the fuel rods are grouped into fuel assemblies holding several hundred of these rods, which can only be moved by a heavy-duty industrial hoist.

"But tell me, why use zirconium for the cladding or sheathing?"

Jake spoke with confidence. "It facilitates the free flux of neutrons between adjacent fuel rods and assemblies."

Enoch stopped and grinned. "Full marks. The fresh fuel assemblies are now ready to be placed in a reactor. But first, some observations about the front end of the fuel cycle. Uranium mines operate in some twenty countries, but more than half of production was sourced from just ten mines in four countries as of 2018."

"Which countries?" Brian asked.

"Australia, Canada, Kazakhstan, and Russia." Enoch resumed his promenade across the room. "Even in the United States, these mines lack sufficient environmental protections, and over fifteen thousand of them are in the western United States. Most of them were stonewalled from having to undergo reclamation and remediation because of the General Mining Law of 1872, and are slowly being eroded by wind and water, with ongoing releases of wind-borne and waterborne radioisotopes."

This summation sparked another entry:

Uranium Enrichment

"The enrichment stage is the most energy-intensive step in the front end, and it's the primary reason nuclear power is not carbon-free as adherents so often claim. Besides the gaseous diffusion associated with Oak Ridge in Tennessee as part of the Manhattan Project, multiple techniques of enrichment are used. Centrifuges are most frequently employed and are at least twenty times more energy-efficient." Enoch clicked his remote.

"Light water reactors, like those used in this country, are designed to offer deep defense against accident or attack, as can be seen in this slide."

Reactor Defenses

- inert ceramic-quality uranium oxide
- airtight zirconium alloy cladding
- reactor pressure vessel made of steel ≥ 12 cm
- pressure-resistant, hermetic containment structure
- security zone around reactor park
- air traffic control zone
- cyber disengagement/protection

After erasing the board, Enoch wrote:

Back End

"On to the back end of the fuel cycle. The dragon in its lair is the spent fuel pool. Remember the conundrum about short-lived and long-lived radioisotopes? The spent fuel assemblies release literally a billion times more becquerels of radiation than the fresh fuel, and far exceed the boiling temperature of water as well.

"A typical pool is forty feet deep, with the bottom fourteen feet consisting of storage racks for the grouped fuel rods. These assemblies have spent three to six years in the reactor pressure vessel, will be stored underwater for at least five years, and then are either reprocessed or installed in above-ground, multi-ton concrete casks for interim storage.

"The water in the pool is both a coolant and a radiation shield. What is referred to as 'radiolysis' is of particular concern, because the radiation flux breaks down the water molecules into their constituent hydrogen and oxygen atoms. This means that the air above the spent fuel pools must be monitored and expelled to avoid buildup toward a hydrogen-oxygen explosion.

"The water is also monitored and circulated. Otherwise, without heat exchange, the water would ferociously boil away and allow the spent fuel rods to melt down and create untold havoc."

Frowning as he faced the board, Enoch jotted down:

Reactor Misadventures

"Nuclear reactors experience a variety of unanticipated disruptions. One of my favorites is swarms, or 'blooms,' of jellyfish. Actually, a better nickname is 'jellies' since they're not really fish. With climate change and its associated marine acidification, common moon jellies and other species of this Cnidaria[12] group are increasing in many coastal areas, uninhibited by ocean acidity.

"In 1989, eighty tons of these cnidarians were removed from a nuclear plant in Madras, India. In 2006, thousands of moon jellies shut down a US nuclear warship docked in Brisbane. In 2011, Scotland, Japan, and Israel all suffered similar outages in commercial reactors. In 2013, a Swedish reactor was shut down for the second time by this same pesky species. Note this list is not even close to exhaustive.

"Another way climate change can interfere with reactor performance is simply through heat waves in fresh water. In 2012, the US saw its first two-week shutdown of the Millstone Power Station in Connecticut for this reason. In August of 2018, France shut down four reactors. In 2019, plants in both France and Germany were curtailed due to heat. Increasing vulnerability with storm surges and sealevel rise are obviously major concerns as well."

Enoch went to work with his marker again:

Pressure Vessel Embrittlement [13]

"Earlier, I mentioned induced radioactivity by neutron activation of the thick carbon-steel reactor pressure vessel (RPV), but coupled with this is structural embrittlement of the metal alloy.

"In 2002, a routine inspection at Ohio's Davis-Besse 3 reactor revealed cracks in the bolted head of the reactor pressure vessel. A corrosive void had almost eroded all the way through, leaving less than a centimeter of metal protecting the core from explosive decompression and generalized radiation release.

"It would have been as disastrous as any of the reactors at Fukushima Daiichi. The NRC pulled their license and the new RPV lid cost the company years of repairs and six hundred million dollars."

Underneath, he added to the list:

Loss of Coolant Accident

"A loss of coolant accident, or LOCA, is another nightmarish scenario. A scram by control rods only stops fissioning in its tracks. Ongoing decay of fission products with releases of alpha, beta, and gamma radiation continue to produce seven to eight percent of the original heat, but if the primary and backup coolant systems fail, the fuel will crescendo to 1,800 degrees Fahrenheit, versus the normal operating temperature of less than 700.

"In this scenario, water vaporizes, pipes burst, fuel rods melt, hydrogen gas is released, and radioactive iodine, krypton, xenon, and cesium are volatilized into the air. And if the configuration regains criticality as it collapses and the molten mass starts uncontrolled fissioning, the rocketing heat causes a slump as hot as lava, which will easily breach the six to seven inches of carbon steel at the base.

"Then, this molten so-called 'corium' reacts violently with the concrete and melts through to the water table and the surrounding outside world."

Turning, Enoch wrote:

Uncontrolled Power Excursion

Enoch let that sink in for a couple of seconds before continuing. "Another event damaging the core is an uncontrolled power excursion. Chernobyl is the primary example. The explosion in 1961 of that experimental reactor in Idaho was a lesser-known and earlier event."

Enoch added the next item:

Haddam Neck Halon

"Here's another classic story," Enoch said.

Jake was curious, not having heard the tale before.

"In 1997, training department staff took a picture of the fire-detection panel at the Haddam Neck nuclear plant in Connecticut. Apparently, the camera's flash tricked the fire detector circuit into sensing fire, and seconds later, the fire-suppression system began discharging bromotrifluoromethane gas, commonly known as halon, into the control room, dropping the oxygen level below that needed to sustain combustion—or life.

"For thirty-five minutes, the operators monitored controls through a window and darted into the control room in response to alarms. I don't know if they used masks with an independent air supply or simply held their breath. All I can say is that fortunately, they didn't abandon ship.

"Reactors kind of need a constant human touch."

Enoch wrote:

Fukushima Daiichi

"Time for a couple of last points. Fukushima Daiichi in 2011 should be instructive to all of us in how quickly a reactor can deteriorate in a 'station blackout'—that is, loss of both on-site and off-site electric power. Reactor 1 melted down first, followed sequentially by reactors 3 and 2 over four days in March. I need to emphasize that seawater from the tsunami did not penetrate far into any of the reactor containment structures, al-though it did badly flood the poorly selected location for the backup diesel generators. Rather, the meltdowns occurred predominantly because of the power failures. Review the ghastly scenario for the first reactor on this slide."

Fukushima Daiichi Timeline — Rapid Progression

- 3 hrs top of core exposure
- 4 hrs core damage
- 5 hrs complete "uncovery"
- 9 hrs corium slump
- 14 hrs melted thru reactor 1, causing violent reaction with concrete

Helicopters made futile attempts to douse flames with borated water.

Bathtub Theory [14]

"Finally, we'll close with the bathtub theory of nuclear plant failures being clustered near the beginning and end of the plants' years in service. See the picture on this slide for reference."

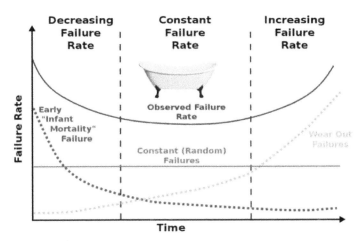

Figure 2: Image by Wyatts (inset photo of bathtub added), licensed under CC BY-SA 4.0

"At the beginning, steep curve 'early infant mortality' failures occur due to material imperfections, assembly errors, and novice worker mistakes during the first several years of operation."

Jake grinned to himself. *Easy to slip in a tub.*

"In midlife, we see an increasing number of refurbishment and replacement events, with occasional unanticipated outcomes.

"Finally, in the last years of reactor performance, 'steep late curve' problems occur as neutron activation and embrittlement cause components to wear out. Clearly, this is the best argument I've heard for not extending licenses past the forty-year design of these plants."

Enoch took a breath and a sip of lemonade. He probably knew his time was up, but Jake saw him check his watch anyway. "Looks like Noah and I are done for the day. We thank you for your interest and attention to this important field."

Jake stood, motioned to both of them, and said, "It is we who must thank you for your review of this 'hot' topic. In both senses, I guess."

"Time to evacuate this particular control room." Jake laughed at Brian's subsequent crack. This was the group's signal to disperse. A few students approached Enoch and Noah to ask lingering questions. Rachel and Brian left hand in hand, while Abbey waited, more or less patiently, for Jake to wrap things up. Pulling on their coats, they headed out into a beautiful late September, conscious of the nip in the early evening air.

NINE

Biotic Resurrection

The fuel of the future is going to come from fruit like that sumac out by the road, or from apples, weeds, sawdust—almost anything. There is fuel in every bit of vegetable matter that can be fermented. There's enough alcohol in one year's yield of an acre of potatoes to drive the machinery necessary to cultivate the fields for a hundred years.
—Henry Ford

It was a Saturday—brilliant sunlight starting the day off enthusiastically, embellished only by a few cirrus clouds. The squad of four had decided to go off campus for their long-planned discussion of biofuels, the only energy sector where no other George Washington faculty members they spoke to had expressed any interest or claimed any expertise. Jake was the de facto leader as usual. Abbey, with her background in biochemistry and growing interest in his pursuits, was a more than willing participant.

Brian Pickrell was Jake's best buddy. Brian was in his last year of a doctorate in mechanical engineering and usually wore his dark brown hair in a ponytail. Originally from Kansas, his undergraduate degree was from the University of Missouri at Columbia, where his major claim to fame was his arrest for being drunk and disorderly. Apparently, his decision to visit a young woman via ladder placed up to the second-floor window of a campus sorority was deemed unwise, particularly given her status as the daughter of a prominent professor in his department major. Jake figured things might have turned out better if his inebriated confederate hadn't broken a couple of first-floor windows trying to take the ladder down after the fact. The story hit the campus and local newspapers over the next couple of days.

Rachel McCready was almost as tall as Brian and wouldn't put up with any of his guff. She had blue eyes and a mass of auburn curls, which she often tried to tuck behind her ears. Her voracious curiosity easily matched his. Hailing from Pittsburgh, she had attended Fordham University and was now in her last year of a master's in international studies and population biology. While not as much of a cyclist as Brian, she had played lacrosse at Fordham—without accruing any record of arrests.

Dragonflies and damselflies displayed their usual acrobatics above several small ponds near Tazza Café and across from the Kennedy Center.

Abbey opted for outside seating since it was such a gorgeous day. With the wide rollout of coronavirus vaccines and marginally sufficient herd immunity reached a few years earlier, the country had stabilized. Infection still infiltrated and smoldered in far too many other countries around the world, a situation which Jake heard her rail against frequently.

Their waitperson, Lyndsey, slim and dressed casually in jeans, arrived to take their orders. Abbey ordered a prosciutto omelet and Jake, corned beef hash, while Brian and Rachel opted for a shared egg-white omelet with a side of hash browns. Each of the four friends had volunteered to split up the task of researching an ancient and entirely different fuel source located within easy reach—past, present, or future—namely: biofuels and biomass.

Jake cleared his throat. "If we're all ready, I'm going to start with a definition of biofuels and biomass that differentiates them from nuclear and fossil fuels."

"Let's do it," Brian said as he sat back, releasing a discreet belch.

"Well," Jake started in, "the conventional concept is mainly liquid fuels sourced from more than eighty percent renewable materials. An example would be E85 for cars, which is eighty-five percent ethanol blended with gasoline as a petrofuel. This requires dedicated flex-fuel vehicles with modified fuel injection and prohibits the use of magnesium, aluminum, or rubber for fuel tanks and engine components."

Brian looked around for their waiter, but she was not in sight.

"Biofuels are derived from contemporary, or geologically *recent*, carbon fixation by dint of photosynthesis in plants and algae, typically biologic material less than a century old, whereas fossil fuels constitute an undisturbed carbon sink, most of which ranges from one to four hundred million years old.

"But biofuels are a relatively labile carbon sink, subject to aerobic decomposition releasing carbon dioxide or anaerobic decomposition releasing methane. Combustion is regarded as equivalent to normal plant decay, except for its higher speed and release of hazardous particulates—"

Jake stopped when Rachel pointed toward the river and called out, "What is that large bird?"

Everyone craned their necks around, and Abbey, who knew her flying critters, said, "Osprey, I think."

Jake watched the bird fly away, and resumed.

"Merged with this effort is the concept of reducing farm operators' fossil fuel consumption as well as avoiding the sacrifice of wild ecosystems for ranching or cropping. And this is just as valid a principle for protecting temperate forest as it is for tropical forest, not to mention peatland, permafrost, and mangrove forest, and so forth.

"Another issue we'll be delving into is seeking drop-in substitutes for aviation fuels. Current aircraft can only be powered by expensive hydrogen or liquid biofuel. I've heard fanciful proposals to bring back dirigibles filled with helium, which is unfortunately nonrenewable. Plus, lighter-than-air zeppelins are bloody slow. Global trade and air travel may necessarily shrink in spite of burgeoning demand.

"However, exciting developments in electric aviation may be competitive for trips up to a thousand miles, and could make flights much quieter for crew, passengers, and communities surrounding airports." Jake saw Rachel's eyes light up.

The coffee and lemonade arrived. Abbey took her first sip as Jake soldiered on.

"Let me tell you about Seattle-based startup Zunum Aero, which had backing from Boeing HorizonX and JetBlue Technology Ventures. Their small regional aircraft are already achieving a seven-hundred-mile range, with the goal of a thousand miles by 2030. The US has over thirteen thousand airports—more than any other country—but one hundred and forty of the largest hubs carry ninety-seven percent of the traffic. Electric airplane companies could inaugurate new routes and connections for five thousand of the smaller airports. The planes could either charge batteries on the tarmac or change out battery packs. The FAA has already developed certification standards."

Rachel broke in. "Are these basically just big drones?"

"Surprisingly not. Most look like actual airplanes. Some of the earliest configurations are plug-in hybrids with a fuel cell backup that carry ten to fifty passengers. With a thousand-mile range, electric airliners could take away at least a third of worldwide air travel from both the oil companies and conventional plane builders. Personally, I think this'll be a much bigger deal than biomass-based aviation fuel.

"And on that positive note, I'm going to turn this over to Rachel to talk about categorization of this energy sector." He finished up with a nod in her direction and a gulp of coffee.

Rachel got started while simultaneously passing out handouts to the group. "Our species, and probably hominins before *Homo sapiens*, have relied on biofuels since first mastering fire. There's paleontological evidence dating back about 350 thousand years of fairly widespread use of controlled fire in Africa and parts of Eurasia. Now, look at the first page of my handout."

Biofuels + Biomass Comprise

- [bio]alcohols: biomethanol, bioethanol, biopropanol, biobutanol
- [bio]methyl esters: biodiesel comparable to petrodiesel
- [bio]mass: woody or other cellulosic material

Finished Products

- collected as methane from landfill or livestock
- synthetic liquid fuels or "synfuels"
- native form converted to chips or pellets
- combusted to generate electrons or battery storage [secondary energy carrier]
- electrolysis to hydrogen gas [secondary energy carrier]

"For broad context, I unearthed several foundational principles." She added sugar to her java. "First, tropical biofuel crops almost always yield more 'caloric,' or energy, content than do temperate crops—by a multiple, not a percentage—partly by benefiting from a continuous growing season. As a pertinent example, the world has four major oil seed crops, namely, canola, oil palm, soybean, and sunflower.[15] If we look at net annual yield or primary productivity per area, oil palms produce eleven times more oil than soybeans."[16]

Eleven, wow, Jake mused.

"Next—and this may surprise you when you think of how much the world relies on wood and other biomass for fuel, including in some pockets in this country—global biomass already provides as much as a fifth of our end-use energy services."

"You mean, like cooking and heating?" Abbey asked.

"Yeah. Finally, liquid fuels are required for most but not all ships, current aircraft, and heavy-duty trucks, which jointly add up to some twenty-eight percent of global transportation fuel demand. Our task today is to determine how much of this requirement might be met by biofuels.[17]

"Clearly, electrification and decarbonization will proceed hand-in-hand. Our civilizational challenge, as Jake likes to say, is squaring the circle and accomplishing carbon neutrality. Now, multiple possible metrics are useful for comparing and contrasting energy crops. But I've decided to boil it down to just three critical parameters.

"So, flip to the second page of the handout."

Analyzing Competing Energy Crops

- sustainable maximum yield per area
- EROEI, or net yield ratio
- inputs + impacts

Yield Per Acre or Hectare

- cubic ft of methane
- cubic ft of hydrogen
- gallons or liters of liquid fuel
- pounds or kilograms of chipped or pelletized fuel
- amp-hours of battery storage, or kWh to grid

"By now, we're comfortable working with the concept of Energy Return on Energy Invested, which is just the ratio of final available energy divided by the energy inputs required to produce it, but I'd actually prefer to call this number a multiple, or perhaps best of all, a *coefficient*. That's it, a Net Yield Coefficient, or NYC."

"I doubt this moniker will catch on, partly because of the confusion with a certain major metropolis," Brian couldn't resist adding. "I love that idea, Rachel, which is to say, 'I heart NYC.'"

"Oh, brother," she muttered.

Then Brian added, "I bet we could sell T-shirts," which earned him a glare and a few chuckles from the other two.

A welcome interruption occurred when their orders arrived. Jake called a halt as everybody had their first well-earned bites.

Rachel pushed ahead. "So, from this point on, I think we can focus most of our analysis on two key ideas: yield per area, plus EROEI.

"But I digress to remind everybody about what we cannot neglect: the other significant inputs and impacts that should factor into our decisions. These are further down the page."

Input & Output Parameters

- soil: types + availability
- water: rainfall or irrigation
- inputs: fertilizers + amendments
- chemicals: pesticides + herbicides

"Don't worry, I won't bring these up again. But you can also see the list of downsides of biofuels." She took another quick bite, licking the corner of her mouth.

Environmental Quandaries

- deforestation + desertification
- vulnerable monocultures
- greenhouse gas [GHG] emissions

Zero-Sum Choices

- food
- feed
- fuel
- feedstock

"This isn't even an exhaustive collection of these problems. And I suspect—belay that—I know people will weigh each of these and other factors differently.

"Pulitzer Prize-winning Michael Ramirez communicated this even better in a great cartoon a couple of years ago, wherein a small dark-skinned kid is holding an empty plate in his lap and a rotund white guy in a suit and tie is standing next to him, each with a hand on one end of an ear of corn. The guy in the suit is saying, 'Excuse me, I'm going to need this to run my car.'

"And we could substitute other crops for that corn: palm oil, peanuts, pecan nuts, potatoes, and pumpkin seeds—all without even leaving the letter 'p.' Almost half of the habitable land surface of the planet is committed to human agriculture." Rachel sat back, looking satisfied, and motioned to Jake.

"Good summation, lady," said Jake. "Next up, Brian is going to drill down a bit into bioalcohols, biodiesels, and direct biomass."

Brian lifted his coffee cup and began. "I don't drink beer at breakfast, but I can tell you that beer is a fermented alcoholic beverage made from water and cereal grains, with a lot of variations on that theme. All the bio-alcohols are not strictly fungible since—"

Abbey interrupted. "Brian, I'm not completely sure I know what 'fungible' means."

Jake jumped in to answer. "Nothing to do with mushrooms. Rather, a term that economists use in referring to exchangeable or substitutable commodities. Though not identical."

"Like beer, or many kinds of coffee?" she said.

Brian said she'd chosen a pertinent example and resumed. "An alcohol is a compound with at least one hydroxyl group—that is, an oxygen firmly bonded to hydrogen on one side and a carbon chain on the other. Methanol, ethanol, propanol, and butanol are the first four alcohols, with, respectively, one, two, three, and four carbon atoms in a chain. And this is my handout," he said, distributing a paper.

Alcohols Long Used as Fuels

- all 4 share high octane rating, burn efficiently since higher ignition temperature
- offsets lower energy density of alcohol fuels vs. petrol fuels
- alcohols by definition contain hydroxyl group

Methanol

- common name "grain alcohol"
- toxic if ingested

Ethanol

- ~95% of biofuel produced in US
- energetics vary with specific plant + process

Propanol [Derived from Propane]

- energy density higher than ethanol
- used in some fuel cell types

Butanol [Derived from Butane]

- energy density closest to gasoline
- less corrosive + water-soluble than ethanol
- distributable via existing fuel infrastructure

E. coli strains have been successfully engineered to produce butanol by modifying amino acid metabolism.

"Biodiesels are not exactly fungible either at thirty-five degrees Fahrenheit, because they all reach the 'cloud point' of early microcrystals at different temperatures, then the even cooler 'pour point' when the fuel filter will be clogged. Peanut oil biodiesel is exceptionally problematic because it 'gels up' at relatively high temperatures, in the upper forties to low fifties."

Rachel said, "Hard to push peanut butter through a straw."

Jake acknowledged the point and jotted notes on his copy of the handout.

Brian continued, "But canola oil, previously called rapeseed oil, usually remains liquid until just above freezing, considered important if you want to start your diesel engine on a cold morning. Safflower oil biodiesel shares this same advantage.

"Woody biomass slated for direct utilization is usually chipped or pelletized. Can be run through what is called a 'hammer mill,' then a subsequent drying operation. Safer than coal for storage since there's no spontaneous combustion. Advantages and sources are listed."

Advantages of Chipped or Pelletized Biofuel

- easy storage + transport
- effective generation of high heat
- resource local + sustainable

Sources of Chipped + Pelletized Biofuel

- wood scrap
- tree prunings
- peanut shells
- rice husks
- sunflower husks

Jake sat forward and frowned. "Isn't this the sort of plant material usually plowed back into the soil to enrich it?"

Brian nodded. "Obviously going to be an issue in common with all these crops, including forestry slash. There's no waste material in ecosystems. We can't harvest for food and fuel without returning micronutrients and macronutrients back to the ground."

"Cattle, for example, are often allowed to graze on the standing stover after the corn harvest, or that strover may be stored in silos for winter feed. In turn, their urine and manure return some of these components to the soil."

A restless silence moved through the group for a few moments until Jake suggested, "Let's push on with Abbey, who's gonna start talking about individual crops appropriate for energy production."

Abbey straightened and looked around the table. "I wanna begin with a couple of perspectives about biofuels." She passed around her outline. "About their sources and physical phases."

Biofuels Come from 4 Sources

- bioethanol from sugars
- biodiesel from oils
- biomass physically processed from plants
- biowaste from forestry slash, agricultural residues, urban recycling

Biofuels Come in 3 Phases

- vaporous biofuel = hydrogen or methane for fuel cell or transport
- liquid biofuel = ethanol or biodiesel
- solid biomass for electricity and/or process heat

Possible to craft alcohol or moonshine out of nearly any plant

---

"Out of scores of potential crops, Jake and I decided we'd limit analysis to six of them: corn—known as maize in the rest of the world—sugarcane, soybeans, oil palm, lignocellulose, and algae.

"Corn-based ethanol will be first, because it has a big, fat target on it. Why? Because we've all been hoodwinked into one of the greatest boondoggles in American history.[18]

"First though, the basics. An ethanol refinery represents a complicated industrial farm project involving about a dozen sequential operations. Two stages are absolutely dependent on microorganisms.

"The first of these uses bioengineered 'amylases'—enzymes which break down the complex carbohydrate of corn seeds into simple sugars. The second is fermentation by living yeast, which convert these sugars into ethanol."

"Corn-derived alcohol is considered a quasi-renewable energy source since it's grown using sunlight—but cultivation and harvesting require significant nonrenewable energy, at least currently. Granted, it also offers useful byproducts, particularly 'distiller's grains' for livestock feed. And the carbon dioxide that corn-derived alcohol produces is marketed for carbonated beverages and dry ice.

"Contrariwise, this form of alcohol creates a ton of harmful effects. Significantly, runoff nitrogen and phosphorus fertilizer create an annual 'dead zone' in the Gulf of Mexico. In 2018, EPA-supported and other scientists mapped out 8,185 square miles of biologic desert in the Gulf. Over four hundred of these episodic, near-shore areas exist worldwide. Finally, US corn production uses the most irrigation water of any of our crops, not to mention half—a full half—of our fertilizer.

"The yield from corn is 494 gallons an acre, which is noteworthy. But we shouldn't be putting ten percent ethanol in our gas tanks, because the EROEI, or net yield ratio, is an abysmal 1.3—an average of fourteen different studies. You've gotta scratch your head at putting almost as much energy into a process as you get out of it. We're wasting billions of dollars spinning our wheels.

"The farm-state senators are extremely powerful, especially in an election year. But pursuing this as a society is irrational." Onlookers at nearby tables glanced over and Brian chuckled out loud.

"Calm down, Abbey, you're getting too loud," Jake said. He put a calming hand on her arm, thinking that an angry pregnant woman was hard to ignore.

Abbey smiled, despite her irritation, took a breath, and continued. "So, the tropical alternative for temperate ethanol is sugarcane; for example, in Brazil and Hawaii. The yield per acre is almost doubled at eight hundred gallons, but more critically, the energy return compared to corn is six for the ethanol alone, but eight if you count combustion of the residual bagasse, the fibrous material left after juice is extracted by crushing the raw plant." She took a deep breath, looking at her notes.

"Biodiesel is next, derived from plant-based oils or lipids, or less wholesomely, from animal fats. Vegetable oil is a liquid at room temperature and has been a part of human cultures for millennia. It's also used for wood finishing, oil painting, and skin care.

"The chemistry of diesel is pretty straightforward. Any diesel fuel is automatically classified as an 'oxygenate,' because it already has a couple of oxygen atoms incorporated into it.

"Gasoline cars and propeller-driven aircraft engines require spark ignition, hence distributors and spark plugs. Diesel engines rely on compression ignition, which allows higher compression ratios and thus more efficient function. Biodiesel burns cleaner than petroleum-based fuel, with fewer exhaust emissions of oxides of sulfur, which produce health risks and acid rain.

"So, here is the whole list of advantages on the next page."

Clean Combustion

- biodiesel [B100] burns cleaner than petroleum-based fuel
- fewer exhaust emission of oxides of sulfur [acid rain]

Other Benefits

- lubrication: pure or blended biodiesel protects engine
- biodegradable + nontoxic in pure form [B100]
- clean odor: in close proximity or urban environment
- energy independence: domestic, renewable resource

"Sadly, the yield per acre is an unimpressive seventy-seven gallons. And the net energy ratio is only about three. Not only are these two metrics pretty mediocre, but soybeans are better used as a premium crop for human consumption and animal feed.

"So, let's look to the tropics to see if there is a superior alternative.

"Oil palm plantations now cover millions of hectares in Malaysia, Indonesia, and Thailand—"

Jake interrupted to ask for a reminder on the size of a hectare, and was told, "not quite two and a half acres." He nodded and took another bite of his corned beef hash.

"Palm oil is an edible vegetable oil derived from the mesocarp, or pulp fruit, of oil palms. It's naturally reddish due to its high beta-carotene content, and grows in forty- to fifty-kilogram clusters.

"The problem is that high-yield trees attract profit-driven investors. The oil palm fruit bunches and shells may also be converted to pelletized biofuel for heating or electricity, so there's an element of cogeneration with this tropical crop, the same point I made about sugarcane." Abbey sipped her lemonade and cleared her throat.

"But oil palm is quite controversial, balancing off food and habitat versus fuel. It's a common cooking ingredient in a tropical belt stretching from Africa through Southeast Asia to Brazil. In Asia, the deforestation or destruction of peat bogs to put in these plantations is associated with emissions of heat-trapping gases. There're also threats to the survival of both orangutans and the Sumatran tiger.

"But the yield in gallons per acre is 635, and the EROEI is a much more impressive eight or nine. Shows again the advantage tropical cultivation has over temperate agriculture.

"Now," Abbey said, "I'll finish by telling you a story about a Beverly Hills plastic surgeon named Craig Alan Bittner. In 2008, it was discovered Bittner was using the fat he'd liposuctioned from his patients to run a home-based biodiesel operation, which fueled his Ford SUV and his girlfriend's Lincoln Navigator. As you might imagine, regulations concerning the handling and disposal of human blood and other tissues are justifiably strict." She was frowning, but a few chuckles erupted around the table.

"The hapless Dr. Bittner lost his license and disappeared into South America, and has not resurfaced so far as I know. Good riddance to bad rubbish, as far as I'm concerned." Abbey lifted her hands and slowly tightened them around an imaginary neck with a growl.

Jake congratulated her with a kiss on the shoulder, and then gestured across the table. "Brian's gonna grab the baton and bring home the anchor leg by talking about an intriguing pair of inedible potential energy crops, one aquatic and one land-based."

But Brian chimed in with a correction. "I'm going to speak first about algae, which we actually do eat; for example, a version of red algae called nori used in sushi rolls."

Jake chuckled and looked to Abbey for help while she shrugged and smiled back.

"Algae are divided into single-celled microalgae and multicelled macroalgae, and almost all of them are aquatic and photosynthetic. Even though they photosynthesize, they're not plants, since they lack specialized reproductive structures, true roots, and leaves. The molecular complexes enabling the photosynthetic conversion of light into carbohydrates are different enough from plants to indicate they split off evolutionarily a long, long time ago. Here's the handout."

Brian's handouts were stained with coffee, but he passed them around unselfconsciously anyway.

Microalgae + Macroalgae

- predominantly aquatic, photosynthetic
- unicellular or multicellular
- ecosystem services
- provide half our oxygen
- food base most aquatic life
- ultimate source crude oil + natural gas [planktonic]
- source food + pharmaceuticals + industrial products

"There are six groups of algae, as you can see on the bottom of the page."

Major Eukaryotic [Nucleated] Algal Groups

- green algae
- red algae
- brown algae
- diatoms
- dinoflagellates

Cyanobacteria 6th Group

- old name "blue-green algae"
- prokaryotic [no nucleus] unlike other algae

When Lyndsey returned to ask about refills of coffee, the group reached a quick consensus they'd all reached their limit and were ready for the bill. "The irony here is that some of these types of pond scum may become the premium source of biofuel in the future," Brian said. "Take a look at the other side of the sheet."

Requirements + Characteristics

- adaptable to saline or brackish water
- algal fuel advantageously high flash point
- harvesting cycle every 1–10 days

Fuel Attributes

- no oxides of sulfur
- reduced carbon monoxide
- lifecycle diminishes CO_2 releases up to 80%

Some species ≥ 60% dry weight oil vs. 2–3% from soybeans

"Algae can only be raised two ways. The controlled, airtight way to raise algae is to pump nutrient-rich water through banks of plastic or borosilicate glass tubes, which results in higher productivity at understandably higher costs. A cheap supply of sterile carbon dioxide will increase yield, though only up to a saturation point. A closed system avoids contamination with bacteria, fungi, viruses, and other algae—but the capital expense is up to a million bucks per acre of greenhouses."

Lyndsey returned with the bill, which Rachel promptly grabbed and handed back to her with her card.

Brian maintained his momentum. "The second way to raise algae is in long, outdoor 'raceway ponds' continuously circulated. By using either technique, up to sixty percent of the dry biomass production consists of oils or lipids,[19] modifiable to biodiesel. As a bonus, any residual carbohydrate can be fermented to bioethanol or biobutanol. But a 'nonfood crop' harvestable within days offers magnificent potential"—Brian raised his eyebrows theatrically at Jake—"yielding an incredible ten thousand gallons per acre and an EROEI of at least five.

"That's the theory, but in reality, scalable, commercially viable systems have yet to emerge. A whole host of problems, as you can see in the handout."

Macroalgae vs. Microalgae vs. Cyanobacteria

- scalable, sustainable, commercially viable systems still theoretic
- bioengineered algae potential invasive species

Vulnerabilities

- wastewater has algal pathogens + predators, metals, chemicals
- bacteria, zooplankton, fungi, rotifers, viruses, protozoans

Requirements for Growth

- water, nitrogen, phosphorous
- efficient production requires concentrated CO_2

Pumping + stirring require energy inputs

"Our final biofuel candidate is lignocellulose from wood—to be more precise, the three main constituent molecules of cellulose, hemicellulose, and lignin. These make up ninety percent of the dry mass of trees. Though counterintuitive, all three of these are polysaccharides—chains of sugars linked together."

Forestry Resource 335 Mt [Million Tonnes] Per Yr.

- logging residues
- urban wood residues
- milling residues
- forest thinning for fire control

Agricultural Resource 907 Mt Dry Biomass Per Yr.

- crop residues
- perennial crops
- grains for biofuels
- processing residues

*Equivalent 2 billion gigajoules or **21%** total current US energy*
But ecologically no such thing as "waste wood"

"Plantation crops of monocultured trees present an alternative to forestry and agricultural residues. Europeans tend to plant short-rotation coppice crops, with sequential harvesting of sectors of the whole plot. They often use poplar, or *Populus*, alternately willow, or *Salix*. Establishment of tree plantations allows fractional harvesting every one to two years. This form of agriculture requires minimal fertilizer and creates less soil depletion.

"In Asia, and to a lesser extent South America, various bamboo species are grown or residues of sugar, rice, and palm oil are used. Eucalyptus trees may be grown even on land with elevated salt levels, and are said to help restore soil for future cereals. Okay . . . done."

Lyndsey came back and Brian asked her to relay their compliments to the chef. Jake leaned back and stretched. "Okay, how do we sum this all up? What do we like? Personally, I relish the idea of electric aircraft. Obviously."

Brian grinned. "Bioalcohol is my favorite, especially in the form of beer."

Abbey rolled her eyes. "Clearly, areal yield and net yield ratio are the two key metrics."

"Corn-based alcohol is an elaborate energy scam," Rachel said, "and the sooner this country comes to its senses, the better. And palm oil farming has terrible habitat and climate costs."

Added Brian: "Algae may become the mainstay of liquid biofuel production, but only in sun-drenched areas of the country, like the Southwest."

Rachel announced the final bill tally while the rest of the group dug into wallets and backpacks to settle their portion.

"Okay, we're copacetic, guys," Jake said. "And I see Brian has volunteered to take home the leftovers."

"My momma didn't raise no fools."

Rachel raised an eyebrow. "Didn't you tell me that you were adopted?" she said, earning her serious tickling until she apologized, still laughing.

TEN

Any Port in a Storm

*The sunsets here were always deep, passionate, and rich—always colors
Camila thought she could take a shovel to and dig at for days.*
—*Kyle Labe,* Butterflies Behind Glass and Other Stories

Bracing is what Rachel called it, and Brian watched her give a quick shiver as the two of them stepped through the doors of District House, heading for the campus dining room on the first floor. They shed scarves and coats at the first empty table, tugging bag lunches out of their backpacks. Rachel dug into a turkey and cheese sandwich, complete with lettuce and condiments, while Brian stuck with his usual peanut butter and blackberry jelly concoction.

Brian spoke between mouthfuls. "Jake was on one of his rants this morning about the grid in Puerto Rico, worried about Hurricane Ricardo on a bearing aiming directly at it. He talked about how the Electric Power Authority failed to learn lessons from Hurricanes Maria and Irma back in 2017, and Fiona in 2022, and that events might follow a similarly sad scenario. He argued there'd been not one but *two* climate signals in the storms in those years."

"What were those signals?" she said.

"The first one is that warming oceans fuel higher wind speeds, the chief but not sole criterion of tropical cyclone 'imperilment,' as Jake puts it, and since 1901, sea surface temperatures have been rising an average of 0.14 degrees Fahrenheit per decade."

"Wait," Rachel said, "that's only about one degree Celsius—and that's enough to turbocharge a several-hundred-mile-wide hurricane?"

"Apparently so, but remember regional warming may be considerably higher than an average, so maybe that's a factor. The second signal is how quickly hurricanes spool up now, with rapid intensification—an increase of thirty knots, or thirty-five miles per hour, over the course of a day. If I remember his numbers, Maria went from Cat. 1 to 5 inside of fifteen hours."

She grimaced. "Are you kidding? That's like sixty miles an hour faster in less than a day."

Brian shrugged. "Irma was also Cat. 5, but I can't remember what he said about how quickly it got worse." He wiped his mouth with the back of his hand.

"People like to deal with single numbers, a single metric. Sure, top sustained wind speeds are damaging and visually arresting on news video, but don't underestimate rainfall, flooding, storm surge, and rip currents."

"Inevitably, a few unfortunate souls get carbon monoxide poisoning from portable generators," Rachel added, both palms up, arms shrugging.

"Anyway," Brian said, "you told me you spent time in Puerto Rico about four months after Fiona. Tell me what that was like."

"That was a couple of years ago, but I'll never forget it. As part of my international studies, I had taken a class on the Caribbean only months before the storm. I petitioned for an independent study experience in Puerto Rico after the storm to figure out why it took so long to get the lights back on. Part of the application process relied on the fact I'd taken two years of Spanish in high school, plus another year as a freshman." She clasped her hands together on the table. "I read up on the Commonwealth of Puerto Rico. The island was first settled by the Taino Amerindians thirty to sixty thousand years ago. The remnants of that culture include words like maracas, hammock, iguana—and of course, *huracán*, or hurricane."

Brian couldn't resist a joke. "So, not just a drink in New Orleans?"

He saw that her hair still glistened from the melted snow. He smiled but she just shook her head.

"Obviously, Brian, Puerto Rico lives smack-dab in the middle of hurricane alley, that band of warm, tropical water stretching from West Africa to the Gulf of Mexico. Because of these storms, they've learned to be more self-reliant. They now encourage urban farming in San Juan and Ponce, and have developed seventy independent and pop-up farmers' markets all around the island," Rachel explained.

"In 1493, Columbus on his second voyage arrived with seventeen ships and maybe fifteen hundred soldiers from Cádiz. They named the large, natural port San Juan de Bautista, in honor of John the Baptist. In 1508, the first European colony, Caparra, was founded by Juan Ponce de León."

"The guy looking for the fountain of youth? I thought that was in Florida."

"Same guy, both places, almost certainly looking for gold, not a fountain. He brought Christianity, cattle, horses, sheep, and the Spanish language."

Brian licked his upper lip. "Probably some diseases as well," he added.

Rachel nodded. "With settlement, they instituted a system called *encomienda*, where native peoples were distributed to Spanish officials to be used as slaves.

"Puerto Rico was a great stepping-off place to both North and South America, so other major European powers made attempts to wrest control of the island from Spain in the sixteenth, seventeenth, and eighteenth centuries. The French and British vied especially to usurp Spanish control.

"Finally, in 1898, the sixteen thousand US troops invaded at Guánica, asserting 'liberation' from Spanish colonial rule. The Spanish-American War ended after only four months, with the American victors granted possession of Puerto Rico, plus Cuba, Guam, and the Philippines.

"Puerto Ricans have had US citizenship since 1917, but residents of the island cannot vote for the president and vice-president or secure voting representation in the US Congress." She put her bag back in her pack.

"Okay now," he said, "what did you learn about the failure of the grid from Fiona?"

"Puerto Rico has 3.3 million citizens, but 1.5 million aggregated electrical customers. And they pay some of the highest electric tariffs in the country with an average residential rate of 23 cents per kilowatt-hour, versus 15.5 cents per kWh in the US as a whole," she pointed out. "In spite of this, almost the entire island suffered a blackout and three-quarters lost access to potable water with Fiona. It was not just the grid—it was also the sewers, streets, highways, bridges, and schools."[20]

"Fifteen months earlier, a company called LUMA had assumed operation of the island's grid, taking over from the allegedly 'corrupt and incompetent' Puerto Rico Electric Power Authority. People said it had been almost impossible to get through to somebody at the utility, even when the grid was up and functioning. Blackouts were common experiences, sometimes monthly. Or worse.

"LUMA cataloged a whole laundry list of defects in the rickety infrastructure they inherited. Thirty percent of electrical substations—for both transmission and distribution—required upgrades just to be safe. Same held true for a full one-fifth of overhead lines. Fifty thousand streetlights were public safety hazards, one outta ten. The whole ground and air fleet used for utility operations did not meet Department of Transportation safety requirements. Vegetation was managed poorly even inside substations, putting overhead power lines at risk in high wind conditions."

"It's tropical," Brian said. "Growing season all year round. I'll bet kudzu grows there, you know, that invasive plant plaguing the South."

"Yeah, I saw a lot of it. LUMA told me they'd gotten federal approval to procure long-lead-time material, like circuit breakers, transformers, and reclosers. And also, money for advanced metering infrastructure and a microgrid project. They talked about a neat idea to support communities with mobile microgrids to fix and reenergize an electrical grid after a major disaster. I heard they were gearing up for thermal imaging and drone surveillance after a big hit like Fiona."

Brian tilted his head back, stretched, and raised both arms. "The proof is not in the pudding but the implementation. The folks managing the grid are clearly unpopular. Jake said after Maria, three hundred thousand people moved to other parts of the US, mostly Florida."

"Voted with their feet, yes they did, Brian. Speaking of which, I gotta go. I've got a class at one o'clock."

Brian reminded her about the upcoming climate lecture in two more days—something about the ocean—and they agreed to meet at Lehman Hall that night.

She stood up, walked around the table, batted her eyelashes at him, and leaned in for a kiss.

Sweet success, thought Brian.

ELEVEN

Holocene Hagiography

Time to get serious about the climate disaster. My children and grandchildren face a planet that is unsustainable otherwise.
—*@Kent, NYT comment section*

The wind had been a constant presence, picking up before dawn's early light. The clouds menaced throughout the day, the sun never piercing through, the gusts playing hopscotch with the fall leaves, tripping about people's feet as they trudged across campus, tangling their scarves, until a sudden, wary calm descended.

Lehman Auditorium had filled up early on this unusually warm, late November evening. Brian and Jake were joined up front by geologist Addie Higgenbotham and a new player, physics instructor Max Baerbock. Extra students and visitors had clearly heard about that night's climate presentation. Jake noticed a couple of newcomers who looked like high school students, probably prospective applicants. Abbey had also caught the bug and had been coming to the sessions that did not conflict with her work schedule.

Eventually, everyone found a seat, leaving Jake alone at the front. He stood next to the lectern waiting for the chatter to die down. "My name is Jake Harper. It gives me great pleasure to introduce our speaker, Mikhail Ligachev from the geology department, who is also a member of the United States Geological Survey. Professor Ligachev is an atmospheric and paleoclimatologic specialist, but is going to lay out the planet's geologic history and the mechanisms controlling its meteorological conditions.

"Feel free to raise questions or offer up comments during the talk. Finally, Professor Ligachev has an astounding amount of information to cover, so his material will be split into several lectures, all of them here in Lehman."

Jake found a seat in the front row next to Abbey as Professor Ligachev, a handsome older man with a full head of more-salt-than-pepper hair, took the floor. He was wearing a nice-looking khaki shirt and pants, finished off with leather boots that had seen some wear. The boots reminded Jake of the oddball character he'd seen at the last lecture during intermission. "Remind me to email Emmanuelle about the sunglasses guy," he whispered to Abbey.

She just nodded in response, seemingly distracted, so he also jotted down a quick reminder to himself.

"Thanks for the introduction, Jake. For those who do not know, Mr. Harper is a newly appointed assistant professor in the engineering department. I've been hearing good reports about these interdisciplinary conferences and about how this student-initiated series has morphed into formal curriculum classes."

Jake tried not to appear too excited, though he knew Abbey would appreciate how important this was to him.

"As Mr. Harper noted, my background is in atmospheric chemistry. To compensate for my bias, I'm going to focus first on the ocean. Notice I said 'ocean' in the singular, because we have only one world ocean which covers seventy-one percent of the planet, an incrementally growing percentage as the sea level rises. After all, aside from isolated bodies of water like the Aral Sea, the Caspian Sea, and the Dead Sea, our globe-straddling ocean interconnects almost everywhere."

Mikhail had already written on the board in the minutes before starting:

One World Ocean

"Let me start with a story about this ocean of ours. In Ferryland, Newfoundland, in 2017—population 465—a bumper crop of icebergs showed up, drifting by and at times grounding outside the intertidal zone. Their height, in some cases, was several hundred feet above the water—Herculean in size. More than six hundred icebergs meandered through the North Atlantic shipping lanes that month, whereas the usual April census is only about eighty. Most of the townspeople appreciated the groundings of these colossal Arctic immigrants as a key tourist attraction.

"These icebergs were, in small part, created by snowfall over Greenland before and since the last glacial period, perhaps fifteen to twenty thousand years ago. Then, glacial calving occurred roughly three years ago in Baffin Bay, and the ice subsequently got caught up in the southbound Labrador Current.

"Big ice is dazzling and extravagant. But, as you might expect, the local fishing industry regarded the arrival of glaciers as a nuisance, and the US Coast Guard, heading up the International Ice Patrol, broadcast recommendations for ships of all types to adjust their routes to a more southerly latitude.

"As an historical aside, the *Titanic* probably sank directly south of Ferryland, not just from executive and command hubris, but likely from ice calved off the west coast of Greenland. But this parade of icy titans several years ago may have signaled, as Sherlock Holmes said, that 'the game is afoot.'

"I've long admired a quotation currently attributed to England's First Earl of Balfour, though oft mistakenly attributed to either Mark Twain or British Prime Minister Benjamin Disraeli. What he famously said in 1892 was this: 'There are three kinds of lies: lies, damned lies, and statistics.' But I'd add graphs to that list. Please look at this slide."

But These Are Not Lies

- Earth colder now than majority prehistory
- Earth mostly ice-free over 4.6 billion years [Ga]
- current planet is the only one known and familiar to us

Assaults on Ocean

- warming
- expansion
- upper stratification
- attrition sea ice
- acidification
- deoxygenation
- cacophonies
- nutrient pollution
- algal blooms
- overfishing
- extinctions
- regional marine heat waves
- coral bleaching
- plastification

"Doesn't anyone want to challenge my claim that the Earth is colder now than it's been for most of geologic history? Here I thought we were all supposed to be having conniption fits about warming, but over a long span of geologic time it was even hotter than it is now. Would any of you object to my assertion that we are basking in an ice age?"

Jake figured he knew what was coming but held silent. Mikhail surveyed the room and saw no takers. He replaced his last heading with a new one:

Ice Ages

"How can I so brazenly claim we're experiencing an ice age? If we postulate that any time that continental areas possess not just multicentennial but multimillenial thick ice sheets in both the Northern and Southern hemispheres, then it's fair to say that a 2.6-million-year ice age—the current Plio-Pleistocene glaciation—is holding sway. Antarctica alone is as vast as the United States and Mexico combined, and has such an ice sheet.

"Any potential confusion may be resolved—as a nontrivial point—by realizing that an ice age is not static but rather consists of alternating glacial and interglacial periods. As the terms imply, a glacial interval is characterized by colder average temperatures, growth of glaciers and ice sheets, and lower sea levels as the mass of frozen water supported by land increases. Each glacial is followed by an interglacial period with warmer temperatures, northward and mountainward retreat of glaciers, and higher sea levels as ice melts into the ocean. Our current Holocene epoch over the last 11,700 years has been just such a 'Goldilocks' interglacial in Earth's history." He paused.

As a great speaker can do so effectively, Jake thought.

"Let's put all this into the perspective of ancient and venerable deep time. The Earth is 4.6 billion years old. Life evolved not quite four billion years ago, but it was microscopic and single-celled for quite a stretch, more than two billion years. It wasn't until the Cambrian period 541 million years ago that significant multicellular life evolved. So, the interval we're discussing all happened in the last one-ninth of planetary history.

"Look at this next slide to visualize the timeline."

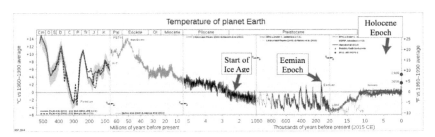

Figure 3: Image by Glen Fergus (modified with three labels), licensed under CC BY-SA 3.0

"You can see," he said, gesturing with his pointer again, "the average surface temperature of Earth over the last 541 million years is graphed beginning with the Cambrian. I should explain that on the left side, units are in millions of years, scaling down on the right side to thousands of years. You can see where our current ice age began, and where the average temperature rose during the interglacial Eemian and Holocene epochs.

"If this all looks complicated, it's because it authentically is. This subset represents only a sampling of the last fraction of all the Earth's ages. Blame the British, because they did so much of the early delimiting of the geologic ages, with hundreds of imaginative Latinate names, but only several of which we need for our purposes.

"Officially, the International Commission on Stratigraphy is the organization that concerns itself with stratigraphical and geochronological matters on a global scale. You've probably seen their periodically summarized charts—laminated, they make great bookmarks—and have perhaps been forced to try to memorize all the fanciful but sometimes genuinely poetic names."

Jake leaned forward in his seat, as fascinated by Mikhail's teaching acumen as by the information he was sharing.

Stepping to the board, Mikhail wrote down four names in chronological order:

Cambrian Period

Cretaceous Period

Eemian Epoch

Holocene Epoch

"The Cambrian period encompasses the flowering of multicellular creatures, beginning, again, over half a billion years ago. Until that point, no life form was larger than a microscopic single cell. The Cretaceous period ended the reign of dinosaurs with a big bang just over sixty-six million years ago when a massive asteroid struck the planet, with the only surviving megafauna being the theropod dinosaurs that evolved into present-day birds. Controversial research suggests that secondarily triggered volcanic activity prolonged the stress on remnant groups of species after that asteroid strike."

Jake was rapidly scribbling notes, as were many in the audience.

"The much more recent Eemian epoch, between about 130,000 and 115,000 years ago, was the last interglacial period before our current blissful Holocene circumstance. We will contrast this with our modern interglacial period in another lecture to see what lessons and warnings may be gleaned."

Abbey was looking over Jake's shoulder, perhaps jealous of his neat cursive—so unlike her medical scribbling.

"Careful inspection of this graph reveals a number of salient observations. A glance confirms that surface temperatures have indeed been higher for the majority of this more-than-half-billion-year recent history of the planet, buttressing my earlier claim. Furthermore, most geologists agree that we have been experiencing an ice age for the last 2.6 million years. But notice there was a switch in amplitude and frequency, beginning about eight hundred thousand years ago, characterized by more regular glacial periods of ice advances interspersed with interglacial periods of ice retreat." Mikhail strode to the left side of the room, leaving a few seconds for everyone to study the slide.

"What is most remarkable is the almost metronomic regularity of glacial-interglacial cycles in this last long interval.

"Why in the world should this be happening in such a periodic cycle?

"A fellow by the name of Milankovitch performed calculations that shed light on the question, but we'll wait until the next meeting to delve into the profound effect of Earth's orbital variations on climate." Mikhail stroked his chin, moving back toward the center.

"Speaking of interglacials, we've been fortunate to live in this Holocene epoch—an extremely auspicious interlude of stability, with a relatively narrow global mean surface temperature spanning less than 0.5 degrees Celsius to a maximum 1.0 degree Celsius variation. Essentially all recorded human history, all systems of writing—and therefore all historical civilizations—occurred during this epoch. Before anyone objects"—he held up a finger noting a hand being raised—"I'm going to gloss over cave art and other artifacts that challenge my thesis here." The hand came down.

"The most recent glacial period peaked as the Last Glacial Maximum, or LGM, between twenty-two and twenty-four thousand years ago, with glacial retreat finally completed roughly by the start of the Holocene. As an identifiable species for somewhere between three hundred thousand and two hundred thousand years, it's clear *Homo sapiens* has survived and persevered through several exhilarating global glacial periods, so perhaps we are a hardy species after all."

Jake whispered to Abbey that he would like to show her something exhilarating when they got home, and she shushed him.

"Almost as remarkable is that through most of the planet's history, if we exclude the initial Hadean eon beginning 4.6 billion years ago, the global mean surface temperature has varied only about fifteen degrees Celsius between upper and lower bounds, a range of merely twenty-seven degrees Fahrenheit. Consider how common it is for us to experience even wider swings in a single day at almost any given location or in any season."

Mikhail paused to write:

James Hansen

"In 1981, Hansen, in conjunction with several other NASA researchers, wrote a seminal article in the journal of *Science* entitled, 'Climate Impact of Increasing Atmospheric Carbon Dioxide,' which sounds pretty prosaic now. But it wasn't until he gave testimony before the Senate Committee on Energy and Natural Resources on a day of sweltering 100.9-degree heat—on August 23, 1988—that he first brought wider public attention to the risks of uncontrolled release of greenhouse gases."

Jake recalled that their fair city had experienced a week above one hundred degrees back in August.

"Aside from water vapor, the three most powerful of these gases are carbon dioxide, methane, and nitrous oxide. This latter culprit is indeed the anesthetic drug used by dentists, though they are not the principal source of its release. Other less recognizable chemicals include hydrofluorocarbons and perfluorocarbons, plus other unpalatable and hard-to-pronounce compounds."

A raised hand caused him to call on Rachel, who said, "Why aren't we talking about water vapor, if it's so potent a heat-trapping gas?"

"A reasonable question," Mikhail replied, "with the only answer being that we live on a water planet and have absolutely no control over evaporation, transpiration by plants, water in volcanic eruptions, and so forth.

Of course, we do have control over combustion, but the main worry is the release of carbon dioxide and methane, not water vapor, which is extremely changeable, or labile.

"Lest you think this is a field of solely cutting-edge, modern science, you should be aware of the investigations of early naturalists. French mathematician and physicist Jean-Baptiste Joseph Fourier scratched his head, back in 1826, when he calculated the planet should be a lot cooler than it was, postulating that the atmosphere was somehow blanketing and protective.

"In 1859, Irishman John Tyndall first proposed that trace atmospheric gases could be the explanation.

"Later, Swede Nobelist in Chemistry Svante Arrhenius demonstrated with benchtop laboratory experiments that carbon dioxide was effective in trapping heat, which he established in—wait for it—1896. So, climate science actually got its start in the nineteenth century."

Jake could only chuckle.

"Switching gears, how much bigger is a square meter than a square yard? A meter is 39.37 inches, so a square meter is about 1.2 square yards. A couple of simple equations are referenced in this slide."

Calculations

- One square meter   =   1.2 sq yd
- Area of circle   = $A = \pi r^2$
- Area of sphere   = $A = 4 \pi r^2$

"Why's this simple algebra important? Because we want to calculate the area of the top of the atmosphere, colloquially known as the 'skin' or TOA. And what is that, you may ask? Atmospheric scientists arbitrarily define the TOA as an imaginary sphere located a hundred kilometers above the surface of our planet, conveniently encompassing practically all of our air—which only seems logical.

"One hundred kilometers is about sixty-two miles, so, pretty high up into space.

"The calculation of the area of a circle is something you learned in middle school. You start with a number identified with the symbol pi, as in apple pie. Some of you may recall pi is approximated as 3.14159. At any rate, multiply this irrational number by the radius of the circle. Twice."

"Voilà. Now you have the area of a circle. And the area of a sphere—such as the TOA—is precisely four times larger.

"The area of the Earth's surface is 510 trillion square meters. The area of the TOA is larger—because it's a bigger sphere—at 526 trillion square meters. But we also need to know the area of the Earth facing the Sun at any given moment. Let me describe both the hard way and the easy way to determine this measurement."

Jake was reassured to still hear some folks typing away.

"The hard way is to resort to second-year calculus. The easy way is to imagine the shadow of the Earth on the opposite side from the Sun, and simply calculate the area of circular shadow."

Mikhail again turned to erase and then write:

Watts Per Second

"Here's our dilemma. Our atmosphere today is trapping a bit more than three extra watts per second of outgoing heat for all 526 trillion square meters of the TOA . . ."

Rachel raised her hand again. "Just to illuminate one point, it's not really the TOA that traps the heat, correct?"

"Good question. No, clearly the TOA is only an abstraction. Greenhouse gases are distributed throughout the atmosphere—mainly of course in the troposphere and stratosphere, the lower two levels—with only an infinitesimal amount of gas detectable by the time you get all the way up to the skin. So the trapping of heat occurs throughout the whole vertical extent."

Abbey followed her friend. "And why's this additional heat measured in watts per second?"

"An even superior question," Mikhail said. "Any power rating is about the rate of work performed per unit of time. So watts per second could be interconverted to joules or calories or even horsepower per second instead. Watts per second, though, is the standard scientific framing of the concept. Absolutely *nothing* to do with electricity in this context.

"A good reference is the Annual Greenhouse Gas Index, which compares the combined warming influence of these gases each year to their influence in 1990, the year the signatories of the UN Kyoto Protocol agreed to make its benchmark. By this point, we are more than fifty percent above that 1990 baseline."

Mikhail again turned to write:

Hiroshimas Per Second

"Back in 1998, someone came up with the idea of communicating more viscerally the importance of the heat from the TOA trapping computation when it was then only a seemingly insignificant 2.29 watts per square meter. Multiplied by the area of the TOA, crunching that number equated to four Hiroshima bombs' worth of heat per second, or about four hundred thousand Hiroshimas a day, mostly ending up in the warming ocean. Nowadays, we are up to about five and a half Hiroshimas a second."

Jake heard a quiet, intense expletive somewhere behind them, and felt the same way.

"Another way to skin this cat is to bear in mind we're raising the surface temperature at the rate of about 0.2 degrees Celsius every ten years. Recall at the beginning of this decade we had reached 1.15 degrees Celsius since the start of the Industrial Age. More worrisome is that this increase is accelerating, and positive feedback mechanisms such as melting of permafrost, methane clathrates, and sea ice may later begin to supercharge this progression. But I digress. Let's save that for later."

Again, Mikhail turned to scribble on the board:

Stratigraphy, or Geologic History

As he wrote, a question came from the back. "How long would the temperature continue to climb if we stopped net releases of greenhouse gases today?"

Mikhail finished writing and turned to face the questioner. "If we magically stopped our net additions of trace gases, over some decades we might see a rise of another several tenths of a degree Celsius from where we are now."

Jake knew everyone immediately suspected the world might not be able to stop at 1.5 degrees unless it brought emissions to a screaming halt now, and also started pulling CO_2 out of the air and putting it underground, into trees, or, counterintuitively, into green cements. The room was silent for a long beat.

"Let's reverse gears and enter a time machine to review the genesis of the planet in its early years. Earth spent about sixty million years in condensation and accretion from a grand protoplanetary disc. The heat of formation and radioactive decay kept it sizzling with intense, early volcanic eruption.

"But as the four inner rocky planets were forming, lots of other objects in the early solar system coalesced as well. This resulted in the age of Late Heavy Bombardment. To get a sense of what happened with this bombardment, just look up at our Moon—either with your naked eye or binoculars or backyard telescope—and assess the profusion of craters, often overlapping, large and small."

Jake whispered to Abbey they should check out the Moon that night.

"I've been instructed to take a break in the middle of these two-hour seminars, so here's a chance to stretch your legs. I'm going to start in with the Keeling Curve of carbon dioxide when we get back." Mikhail stepped to the side of the lectern, and immediately several students approached him with questions while most folks headed outside—possibly to stare at the Moon and absorb Mikhail's eloquent words.

TWELVE

Call to Duty

You cannot stop people from dying.
The most you can do is stop them from dying today.
—Anonymous

J ake and Abbey mingled with the crowd making its way outside. Brian and Rachel joined them on the steps. The breeze was the lightest of zephyrs, chilly now that the sun was down, but the stars remained bashful, unescorted by any moon.

Rachel put her arms around Brian from behind, and he covered her arms with his. Abbey was showing more than a bump, and Jake told the unbelievable story of having been first to feel the baby move only a couple of days ago, which Abbey explained was called "quickening," an archaic, yet still useful term.

She wasn't at all embarrassed that she hadn't felt the baby move first. "Jake just happened to be in the right place at the right time," she said.

Brian raised his brow. "Just as he was at the very beginning of this pregnancy." Only he and Jake laughed.

"We've seen our certified nurse midwife for four visits so far," Abbey said, ignoring his comment. "Her name is Charly, short for Charlene, and she is quite young. As you guys know, I'm going to deliver here in the med center at GW. You also know I had my first sono last week, and the blood tests and everything look fine."

Rachel chuckled. "But you wanted to know what gender the baby was and Jake didn't. So how're you gonna keep that secret for another five months?" She squinted. "You can give me a hint later, okay?"

Abbey just laughed. "Benjamin Franklin said three people can keep a secret if two of them are dead. I'm not threatening you, understand, just quoting a reliable historical figure." She reached into her pocket as her phone vibrated and stepped away to take the call.

"Yes . . . yes . . . another patient with diffuse bleeding and history of injection drug use. Admitted yesterday morning to what floor?" She paused, listening intently. "Transferred last night to which ICU bed?"

"Who's the resident on the case?" Again, a hiatus. "I'm assuming orders have been written for both contact and respiratory precautions . . . have them get serum for unusual viruses. We may need to consider hemorrhagic fever viruses again this time. I wouldn't even have put that on the differential if we hadn't had that case in the ER last month." She switched the phone to her other ear. "I'll be over when we finish the lecture. Okay, thanks."

Walking toward the others, she said, "I've gotta go in after the lecture finishes. So I'll see you back home later. Unusual case in the ICU. Jake, I'll see you later, honey." Jake didn't tell her to be careful. The two of them had already talked about the risks of her profession, and he knew she'd suit up in the appropriate protective equipment.

Which didn't mean he couldn't worry.

THIRTEEN

Restless Planet

J ake put aside his concern and called for quiet. Mikhail picked up right where he had stopped less than a quarter hour earlier. On the board, he had scrawled:

Keeling Curve

"A great place to start is with the Keeling Curve of atmospheric carbon dioxide monitoring, which you see on this first slide. In 1958, David Keeling, working under the auspices of UC San Diego, set up a monitoring station near the top of Mauna Loa on the Big Island of Hawaii. The site was chosen because it's in the middle of the Pacific and far away from sources of combustion, biofuels, biomass, or fossil fuels. And most of the time far enough away from volcanic eruptions, except for 1984 and 2022. With a brief hiatus from federal budget cuts in 1964, a 124-foot aluminum tower has been collecting CO_2 measurements more than two miles above sea level nearly every hour, every day, for almost sixty years. David's son Ralph Keeling is carrying on his father's project."

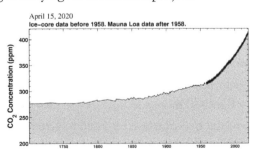

Figure 4: Image by Scripps Institution of Oceanography at UC San Diego, licensed under CC BY 4.0

"The early work was critical since it was not duplicated elsewhere, but now there are hundreds of other monitoring stations around the world, including more than seventy operated by NOAA." He took a moment to make sure the remote was working, then traced the curve upward with his laser.

"Preindustrial carbon dioxide levels were about 280 parts per million. We have records from glacial cores in Greenland going back six hundred thousand years and Antarctica for eight hundred thousand years, so we can tell you unequivocally we're higher now. From other proxy data, we can state that carbon dioxide now surpasses any value in the last three million years.

"The majority of the planet's landmass is in the Northern Hemisphere, which logically supports the lion's share of all vegetation. Carbon dioxide hit four hundred parts per million during a May peak in 2013 when all that exuberant growth was taking off in the North and extracting it from the air with the onset of spring growth. At the beginning of this decade, we had progressed to somewhere in the neighborhood of 417. You can monitor all such vital signs at NASA's Global Climate Change website.

"Carbon dioxide is a persistent gas. A quarter of released CO_2 may still be in the atmosphere and ocean up to a thousand years later. The synthesis of carbon dioxide takes but a moment in an exothermic reaction; for instance, the moment you turn the key in the ignition of your car. 'Exothermic' means heat produced via chemical reaction of burning gasoline or diesel, which drives the pistons of the engine with rapidly expanding gases. The quantitative relationship, or 'stoichiometry,' of combusting gasoline is that one pound of this fuel is converted into three pounds of carbon dioxide. Or one gallon of gas is converted to twenty pounds of carbon dioxide.

"But the truly remarkable fact is that, over centuries, a single molecule of CO_2 will trap a hundred thousand times more heat as a greenhouse gas than the burning of the hydrocarbon required to generate it. In a nutshell, this is why it's so critical to electrify transportation, not to mention develop bike lanes, mass transit, and other forms of urban redesign."

Again, Mikhail took a moment to write on the board:

Ocean Acidification

"Mounting acid levels in the ocean have been called the 'equally evil twin' to global warming. The ocean at equilibrium holds a store or sink of fifty times more carbon than the atmosphere."

Fifty . . . more than I would have guessed, Jake thought, biting his lower lip.

"When a molecule of carbon dioxide enters the ocean, it combines with a water molecule to create carbonic acid. Same as when a bottling plant injects carbon dioxide into a can of soda. If you then separate off a pair of hydrogen atoms, you are left with a carbonate ion. Using this carbonate, mollusks and many other creatures are able to synthesize their shells. Now, check out this next slide."

Ocean pH or Acidity
- baseline about 8.1, on the higher pH, or basic, side since neutral is 7
- coral bleaching in areas with decline to 7.6
- drop of 0.1 unit since year 2000, progressively more acidic
- projected drop of 0.4 units by end-century

"The range of pH is from one to fourteen, with one being extremely acidic, like undiluted hydrochloric acid, seven in the middle being neutral, and fourteen being extremely basic, like strong lye. The human body must maintain its own acid-base balance in a very narrow range around 7.4, slightly on the basic side. Reaching a pH of 7.0 sounds neutral but is incompatible with life—"

Abbey raised her hand instantly. "Mr. Ligachev, I have to comment. As a doctor, I've got to say 'reaching a pH of 7.0' usually occurs in the middle of a resuscitation that's not going well. But technically, some patients can be salvaged, even from a value as low as 6.8—aside from a few more extraordinary cases. I have accomplished that myself, in fact."

With a short bow, Mikhail said, "I stand corrected then, thank you. As far as the ocean goes, however, though there's a lot of spatial and day-to-night variation, the pH should average about 8.1, quite a bit more basic. Now is that an unsuitable blood pH for a person to survive, would you say?"

"Absolutely, nothing above 7.8 would be endurable," she said, and Jake thought she positively glowed.

Mikhail beamed. "As carbon dioxide dissolves into water, carbonic acid increases ocean acidity, lowering the pH. With the change in equilibrium, carbonate ion is diminished. As I previously stated, a variety of sea creatures use this for the calcification of either an exterior shell—an exoskeleton—or interior skeleton constructed predominantly with calcium and carbonate or 'calcium carbonate.'"

Mikhail scrounged around for an eraser and expunged most of what he had written, replacing it with:

First Climate Control Knob

"I'll summarize by saying everything I have explained so far constitutes a description of the first of four main climate control knobs, namely heat-trapping, or greenhouse gases.

"Richard Alley is an extraordinarily articulate professor at the University of Pennsylvania with about forty years of glaciology experience, largely in Greenland, and I believe he may have championed if not inaugurated the phrase 'climate control knobs.' Now, here is the next control knob."

Then Mikhail wrote:

Second Control Knob

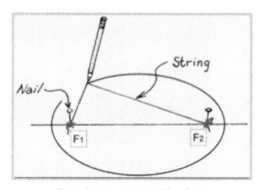

Figure 5: Image courtesy of Doc Green

"Isaac Newton developed, probably concurrently with his rival Leibnitz, calculus—partly to investigate planetary orbits. One insight available to Newton was Kepler's observation that, when plotted on paper, orbits with constant time intervals swept out an identical area bounded by the Sun, indicating the planets were moving faster when closer to our own star."

He stopped speaking to draw an ellipse on the whiteboard with a point labeled inside each end, F1 and F2, just as on the accompanying slide. "If we were to place a small rubber suction cup on top of each focus on this whiteboard, I could attach a string longer than the distance between the two points. If I held a pencil tautly, or in this case, a marker pen, against the string, you can see how I could draw out a nice ellipse. The obvious inference is that each point on the ellipse is the exact same combined distance from the two focal points. Celestial mechanics may be complex, but the underlying math is beautiful.

"Perihelion is the distance of nearest approach to the Sun, and aphelion, the farthest point. Mathematically, an orbital ellipse is defined by two objects orbiting around each other, though more precisely, each is actually orbiting around the changing center of mass between them.

"Everybody thinks they know the Earth is ninety-three million miles from the Sun, but this is only the average value. At perihelion, we're about 91.5 million miles away, and at aphelion, some 94.5 million miles out."

Under his previous notation, he added:

Milutin Milankovitch

"Now, I have an inspiring story to tell you, finally, about a young fellow named Milutin. Born in 1879, Milankovitch was a Serbian mathematician, astrophysicist, and climatologist. He worked in his early years as a civil engineer—building dams, bridges, and aqueducts in reinforced concrete throughout Austria-Hungary.

"Milankovitch was later offered the chair of applied mathematics, celestial mechanics, and theoretical physics at the University of Belgrade. In that era, astronomer John Herschel, geologist Louis Agassiz, and others were exploring the possibility of astronomically-related, cyclical modifications of climate.

"Milankovitch was encouraged to calculate the variations in 'insolation' or solar flux, at fifty-five, sixty, and sixty-five degrees north latitude, congruent with the consensus that summertime heating of the surface of the far North, or boreal Northern Hemisphere, determined the advance and retreat of glaciers at those latitudes. On this slide, look at these three classic cycles Milankovitch incorporated into his theory of the influence of orbital mechanics."

Celestial Mechanics or Milankovitch Cycles

- eccentricity 100,000-yr. cycle described by Johannes Kepler 1609 CE
- obliquity 41,000-year cycle between 22.1° + 24.5°
- precession 23,000-year cycle discovered by Hipparchus in 130 BCE

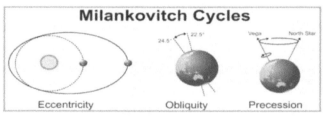

Figure 6: Figure 6. Image by John Cook, licensed under CC BY 3.0

"I will immediately emphasize that the two extremes of the eccentricity cycle on the left are vastly exaggerated. But, look at the summaries of all three of these cycles on the next slide."

Eccentricity or "Ellipticity" Cycle

- ~100,000-yr. cycle
- result gravity of Jupiter + Saturn with different orbital periods
- southern hemisphere 6% more solar energy when closer to "circular"
- current ellipse now getting compact, closer almost to "circle"

Obliquity, or "Axial Tilt" Cycle

- ~41,000-yr. cycle between 22.1° + 24.5°, currently at 23.4°
- tilt result of large collision early in life with moon-forming planetoid
- greater tilt warms both poles + favors deglaciation
- guess which way we're headed now?

Precession, or "Wobble" Cycle

- ~23,000-yr. cycle
- product tidal forces Sun + Moon
- wobbles like a top spinning slightly off-center
- Polaris our pole star now, but in 13,000 yrs. will be Vega

"Why are we concerned about this complex set of orbital influences on our planet? Because the formation or melting of ice in the vicinity of sixty-five degrees north latitude turns out to be a major driver of our fluctuating climate. And these variations in Earth's orbit constitute a potent forcing toward climate change between glacial and interglacial episodes over tens of thousands of years. Note that Milankovitch was the chief architect of our early understanding.

"Recall that the half of the globe facing the Sun is receiving energy at the top of the atmosphere, or TOA, in an area exactly the size of Earth's shadow—but on the side facing the Sun. Then pay attention to the changes in greenhouse gases trapping a seemingly innocuous extra three watts per square meter—but over the area of the entire TOA, which is precisely four times larger."

Mikhail's voice grew more intense, and he made fists with both hands. "Here are the brass tacks. In equilibrium, the heat escaping the planet has to squarely equal the heat being absorbed by the planet. If less heat is exiting, then the Earth heats up to cause more infrared to be directed outward. Thus, a hotter equilibrium is established. It's really that simple. Now, look at this slide."

Variation of Absorbed Sunlight with Combined Cycles

- ~1366 Wm-2, or watts per square meter, arrive [in area-of-shadow equivalent]
- 30% immediately reflected back out from clouds + surface
- 70%, or ~956 Wm-2, therefore absorbed in atmosphere, land, and upper ocean
- incredible ~20 Wm-2 variation resulting from solar ice + snow melting at 65° N

Human-caused Perturbations

- heat as infrared radiation trying to escape in all directions day + night
- TOA is 4 times larger area than Earth's shadow
- sum of average escape must equal ~956 Wm-2 or planetary temperature will change
- extra 3 Wm-2 trapped due to deforestation, cement, fossil fuels [FF 88%]

Jake stopped writing, trying for a moment to wrap his head around the idea that orbital cycles mainly changed the amount of sunlight hitting the crucial band of ice and snow centered around sixty-five degrees north latitude, which changed albedo—the ability of surfaces to reflect heat—and raised the sea level enough to lead to a huge forcing, which is what scientists call any influence changing the climate, of twenty watts per square meter. He thought to himself, *These average twenty watts are trapped day and night everywhere around the globe, and the TOA is four times bigger than Earth's shadow.*

Brian spoke up, interrupting Jake's train of thought. "Where are we currently in the eccentricity cycle?"

"We're actually fairly close to the most distant point and gradually decreasing, and will return to this point in a hundred thousand years," Mikhail replied. "For now, on or near January 3 each year—at perihelion, or closest approach to the Sun—we are nearly three million miles closer than on or near July 4, which is aphelion, or farthest distance from the Sun.

"That's a difference of about 3.4 percent in distance and 6.8 percent in incoming radiation."

Brian responded, "You're saying Australia right now—the whole Southern Hemisphere—gets almost seven percent more intense sunlight in their summer than we get in ours?"

"Absolutely. Might explain the higher incidence of skin cancers in Australia. But due to Milankovitch's third cycle—namely axial precession or 'wobble'—in about thirteen thousand years, conditions will have flipped, with the Northern Hemisphere experiencing more extremes in solar radiation and the Southern Hemisphere more moderate seasonal variations.

"Precession also gradually changes the onset of the seasons, becoming earlier over time. It also adjusts the pole stars to which the planetary axis points. Currently, we have Polaris for the Northern Hemisphere, but in thirteen thousand years, it'll be Vega instead."

"Could you back up and explain radiative forcing again please?" Rachel requested.

Mikhail replied immediately. "The pertinent example is simply solar heat trapped by excess greenhouse gases. Climatologists call this sort of thing a 'forcing.' Perhaps 'driver,' 'impact,' or 'influence' would have been better terms, but 'forcing' is absolutely fixed in all the literature now."

Brian spoke up. "Under the obliquity cycle on that first slide, it states greater tilt warms the poles and melts ice, and that currently we are at 23.4 degrees of tilt. Since we're getting hotter, are we assuming we're heading in the direction of greater tilt?"

Mikhail looked bemused. "I put that on the slide as a trick question, really. No, on the contrary: We're actually moving toward a more vertical orientation, which means our explosive pace of adding greenhouse gases is out-forcing the obliquity cycle." He stopped for a moment to let his words sink in.

"Milankovitch knew he was not dealing with a two-body problem in orbital mechanics, but rather, a hand of poker, the complexity of a five-body system. Sun, Jupiter, Saturn, Earth, and Moon. And there were no computers back in those days.

"The professor decided that with all his teaching responsibilities, he could only devote four morning hours on two days of the week and ten on Saturday to perform this mathematical analysis. Sunday, of course, was for family and church. But then he performed a calculation of his calculation and figured out it was going to take him twenty years to complete this research—which apparently did not deter him from getting started.

"Fate intervened, in the winsome form of Kristina Topuzovich, as in any good, romantic tale. After their wedding, the happy couple left for their honeymoon in Austria-Hungary. But fate has twists and turns, apparently, as World War I broke out, and because he was a citizen in the Serbian military reserves, Milutin was bewildered to find himself a prisoner of war.

"Immediately, his new wife went to work, and through social connections achieved his release under the parole conditions that he not perform any war-related work and that he spend his entire captivity confined to Budapest. The Library of the Hungarian Academy of Sciences took him in eagerly and enabled him to work undisturbed for four years of the war, greatly advancing his mathematical progress.

"He was able to formulate a precise, quantitative climatological model for both reconstruction of the past and projection into the future. The combined effects of overlapping orbital eccentricity, obliquity, or tilt, and precession, or wobble, produced a signal that explained the regular periodicity of glacial episodes. And the signal is a potent one, equivalent to an additional twenty watts per square meter at the top of the atmosphere, and the sixty-five-degree latitude is the chosen parameter climate scientists use to this day.

"So, the Milankovitch cycles—and this is important—are all about a redistribution of solar energy over the planet. Since neither the production nor the trapping of energy is involved, this is not a direct forcing but rather an indirect influence, uncovering darker land and ocean waters which capture and hold on to infrared, or heat, energy. His laser focus was not on the entire globe, but rather the vulnerable high latitudes where glaciers and ice sheets are susceptible to ice accumulation or attrition, located in both the Northern and Southern hemispheres, and first and foremost, the critical northern tier.

"Compare this to the heat-trapping effect of extra greenhouse gases at the beginning of this decade, namely three watts per square meter, as discussed earlier. Greenhouse gases are constantly trapping these three watts over every square meter of the skin at the top of the sky around the entire globe.

"We've now achieved an understanding of the second major control knob of our climate after greenhouse gases. Heat-trapping gases first. Milankovitch cycles second. I'll bet you can't guess what comes in third place, not in order of power or importance, but instead in its more protracted effects.

"If this is any consolation, celestial mechanics—the music of the spheres—is something humanity can't screw up even if we wanted to." That got him a laugh, during which he erased what he had put up and replaced it with:

Carbonate Rocks

"The third major control knob is one where humans may actually have limited influence. This is called 'weathering' of carbonate rock, or our 'mountain thermostat.' It's not only slower than greenhouse gases, but even slower than orbital cycles. Weathering nevertheless offers a potent control. I don't say 'glacially slow' anymore, because glaciers are now accelerating their push toward the ocean almost everywhere.

"Tropical mountains constitute a patient and unrestrainable carbon sink. Here is the sequence: Tectonic plate collisions result in 'orogeny,' or coastal mountain formations like the Cascades in the Pacific Northwest. It takes millions of years to uplift oceanic crust, but since it is formed from mid-oceanic magmatic material, it contains igneous rock rich in magnesium and calcium, which are important here. Again, some of this material is organized on this slide."

Tropical Uplift as Third Major Control Knob

Minerals Combining with CO_2 to Form Carbonate Rocks

- calcium oxide or quicklime
- serpentinite
- olivine
- magnesium silicate hydroxide

Temperature Matters

- Siberian Traps = volcanic eruptions 252 million yrs. ago
- rich in carbonates, but Siberia too cold to absorb much CO_2

Precipitation Matters

- Saudi Arabia has heat + rocks galore
- but no rain to expose + erode rocks

Tropical mountain-building collisions coincide with nearly all of the half-dozen ice ages in the past 500 million years

---

"More broadly, it's hypothesized that low-latitude tectonic 'arc collisions' drive cooling by uplifting volcanic rock—types called 'mafic' and 'ultramafic'—from oceanic crust and the underlying upper mantle of the Earth in the warm, wet tropics, where the rock combines with carbon dioxide by chemical weathering. This igneous rock, such as serpentinite or olivine, is rich in calcium and magnesium which react with atmospheric carbon and bind and sequester it for millions of years. A potent but plodding principal control knob of climate, far slower than paint drying."

Way, way slower, thought Jake.

Mikhail wrote down the next topic:

Tectonic Plate Action as Fourth Major Control Knob

"Our fourth and final major control is notable for one commonality and one distinction. It shares with orbital variations the lack of any human responsibility. Despite my favorite bumper sticker, 'Stop Continental Drift,' influencing it, thank goodness, is not really within our power. Its dissimilarity from other climate control knobs is that it must be the slowest of the controls, taking hundreds of millions of years to effect change.

"Perhaps an example is in order." He set the remote down on the lectern.

"Central America did not always connect North and South America. A lot of it is mountainous, being part of the Pacific Ring of Fire, but there used to be a gap at the level of central Panama. The 'Panama Hypothesis' describes the gradual closure of the Panama Seaway, between 13 and 2.6 million years ago. This had far-reaching ramifications, including but not limited to decreased mixing of the Atlantic and Pacific water masses, accentuation of the Atlantic thermohaline circulation, and more precipitation in northern high latitudes. This gestalt culminated in intensified Northern Hemispheric glaciation during the Pliocene between 3.2 and 2.7 million years ago.

"After all, mighty glaciers are formed by the gradual accumulation of delicate snowflakes."

There he goes again, waxing philosophic, Jake thought.

"All this was accomplished by tectonic plate motion triggering a cluster of volcanic eruptions that obstructed the geology of that Panamanian Seaway, and then marine currents brought in deposits of sediments that closed it off altogether and dramatically altered the ocean and the global climate regime. The arrows on this map mark where the volcanism and sedimentation obliterated the gap. Until this channel was closed approximately six million years ago, the Atlantic Ocean flowed east to west into the Pacific Ocean. But once the isthmus was constructed, currents were dramatically reshuffled."

Until we built the Panama Canal, Jake thought, then remembered the locks.

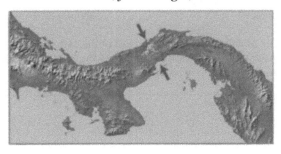

Figure 7: Image by Sadalmelik (arrows added), licensed under CC BY-SA 3.0

"To give you a sense of how this changed the global climate, recall the Gulf Stream, part and parcel of the Atlantic Meridional Overturning Circulation, the famous AMOC. Spanish explorer Juan Ponce de León first noted the Gulf Stream, then Benjamin Franklin later mapped and named it with temperature measurements on a voyage from England to North America in the eighteenth century.

"We now know the Gulf Stream carries about thirty times more water than all the rivers of the world combined. And this current brings enough heat from equatorial West Africa and the shallow bowl of the Caribbean all the way to Iceland and western Europe, such that they enjoy a maritime climate twenty degrees Fahrenheit warmer than they would otherwise. It's estimated that 'Scandinavia receives a heat equivalent to seventy-eight thousand times its current human energy use.' This is a heat pump on a planetary scale, as Professors Sorrentino and Apfelbaum discussed with you some weeks ago."

Mikhail picked up the remote and turned to face the screen. "There are other controls on climate—it's a very complex field—but I've summarized some of the most powerful mechanisms on this slide."

Climate Control Knobs

Assemblage of Controls	Timeline
atmospheric heat-trapping gases	decades to centuries
orbital variations	tens of thousands of yrs.
rock weathering	millions of yrs.
tectonic plate motion	hundreds of millions of yrs.
bolide collisions [comets, asteroids]	days to weeks
volcanic eruptions, including subsea	months to yrs.
biologic evolution	hundreds of millions of yrs.

"Since the rise of industry in the second half of the 19th century, when widespread emissions of greenhouse gases began, the world warmed by about 1.15° C at the beginning of this decade."

Mikhail turned to write one last time:

Summary

"I see the mood here has gotten a bit somber. I realize that resolving our climate predicaments seems like an almost insurmountable challenge. The Holocene is the only home we've ever known. But be of good cheer, for it's my understanding you've discussed in the other sessions some of the energy systems that will ride to the rescue. Energy efficiency is always front and center. Wind. Solar. Hydropower. Geothermal. Biofuels. Tidal power. Wave power. Heat pumps and induction cooktops. Electrification of transportation.

"So, see you all next time." And finally, robust applause erupted as he set down the remote and marker pen on the lectern.

Jake stood and reminded everyone that the next meeting would be in the same room. "Hang in there, folks. This story has a wonderful ending."

Some large snowflakes were just beginning to fall as Jake walked out with Brian and Rachel after a quick hug goodbye with Abbey. He stared down at his feet and nobody talked much. At varying intervals, all three of them turned up their collars.

FOURTEEN

Encroaching Plague

There have been as many plagues in the world as there have been wars, yet plagues and wars always find people equally unprepared.
—Albert Camus

Abbey had not met resident Peter Flanagan previously, and she therefore introduced herself to get the report. He had just meticulously removed his gloves, washed his hands with chlorhexidine, and picked up an alcohol wipe to clean off the bell and diaphragm of his stethoscope. They were standing outside the ICU room with its sliding glass door closed because of respiratory precautions. Peter was wide awake as it wasn't even close to ten o'clock at night, but Abbey knew he'd look worn by morning report. He appeared not to have shaved for a couple of days, his reddish hair mussed up from the ties of the mask he'd now dropped down around his neck. The ICU was full of the background beeps and chimes of alarms requiring nursing attention.

Peter began as requested. "Tyler Carrington is a twenty-seven-year-old white male admitted for the first time to this hospital yesterday morning, brought in by friends who dropped him off in the ER and—"

"Do we have the contact information for those people?" Abbey interrupted. "Have you notified public health and explained how important it is to surveille and test them?"

He looked past her shoulder and nodded in response to both questions. "Tyler suddenly became ill in the morning three days before admission, with chills and unrecorded fever and sweats, throbbing, bifrontal headache, back pain, joint pain, stomach pain, and extreme fatigue.

As an injection drug user, he was initially reluctant to let his friends bring him in, but they finally persuaded him because he couldn't stop vomiting."

Abbey watched his face intently, not taking notes since she'd be reviewing the chart in detail shortly. "No hematemesis, no coffee-ground material, no bilious vomiting, no diarrhea?"

Peter was shaking his head. "No, and in fact once we got him rehydrated and his electrolytes corrected, his upchucking pretty much became a nonproblem. When Tyler arrived in the ER, he denied any travel history or tick exposure but had obvious tracks, not just on his extremities, but also over his jugular veins. My intern thought he also had them under his tongue. He was febrile then, and since admission, has ranged from thirty-nine to forty-one degrees Celsius. On initial exam, I noted conjunctival injection, a flushed face, and a red throat without exudate together with small red petechial spots on his hard and soft palate. But he was not yet clinically jaundiced."

Her eyebrows rose. "Implying he's jaundiced now?"

"Total bilirubin is up to 7.5 and his alkaline phosphatase is twice the upper limit of normal. He has antibodies for hepatitis C virus, of course, like almost all of these patients, but fortunately, he is negative for markers of hepatitis A and B. We'll defer vaccines for those two until discharge."

Abbey knew that while the patient was this ill, his immunologic response to vaccines would be suboptimal. *Odds are he may not make it to discharge, anyway.*

"Lungs were clear to auscultation and percussion, heart without gallop or murmur or rub," Peter said. "Because he is a self-injector, I categorized him as a shooter with a fever, and we got multiple blood cultures and eventually resorted to a central line when we failed to locate a good site for a peripheral IV. Started him on vancomycin and gentamicin since he reported putative penicillin allergy. Echocardiogram showed no signs of endocarditis, and four blood cultures are still negative as of this afternoon.

"The reason I suggested an ID consult to my intern is that he's showing no focus of bacterial infection—and now he's bleeding," he added.

"From multiple sites, in fact," he continued, "even around the central line dressing. Petechiae are arriving in showers on various body surfaces. He's confused and combative, even though a lumbar puncture was negative on the first day. His head CT was negative for abscess or bleeding, but his spinal tap site is oozing even with a pressure dressing."

"You can't tap him again, now that he's a bleeding risk," Abbey said.

He shrugged. "Initially, we had to make sure he didn't have meningitis. His coagulation parameters were okay at the time of the lumbar puncture. Platelets down only a little at 115,000. As far as coagulation tests, prothrombin time and partial thromboplastin time were within normal limits. But that was several days ago. Since then, platelets have dropped as low as fifteen thousand and his prothrombin time is prolonged a couple of seconds."

"You're replacing those blood components?" Abbey said.

"With either fresh-frozen plasma or cryoprecipitate for the coagulation factors. Ten units of platelets at a time seem to suffice once or twice a day to bring the count up to eighty thousand or so. He's received a couple of units of packed red cells on . . . four occasions, I believe."

Abbey pointed out the obvious. "Viral titers aren't back yet or you'd have mentioned that already. Since he's still in the acute phase of the illness, you could use that early serum—assuming you had the lab save some—to try to detect the set of viral hemorrhagic fever antigens by ELISA or viral RNA by PCR tests. Later in the course of the disease, he will develop antibodies. But probably not yet." She knit her brows. "Given the other recent case, Crimean-Congo hemorrhagic fever virus is my first thought, but Marburg, Ebola, and the rest are possibilities."

"Okay, yeah, the lab has some saved serum," Peter said. "So, we had a couple of key questions for you. Given the patient's current status—and since I understand mortality varies widely, from three to forty percent—what are his odds? Second, what do you think about empiric ribavirin? Third, should we hold off on notifying the CDC?"

Abbey let his questions hang in the air for several seconds. "I need to review the chart and examine the patient before I can answer the first question. But you have a jaundiced patient with bleeding and delirium, so clearly his prognosis is bleaker.

"As far as the antiviral agent ribavirin goes," she pointed out, "the best data I found after that last patient was a Cochrane Review from the NIH, a meta-analysis from 2018. They only found one randomized controlled trial with 136 patients and four nonrandomized studies with 612 patients. Four of these five studies were from Turkey, one from Iran. The Cochrane consensus was that we simply don't know if ribavirin reduces mortality, length of hospitalization, or requirement for platelet transfusion. The CDC is still telling us there's no well-proven vaccine or any effective drugs.

"Ribavirin is not without its risks, anyway. It can cause hemolysis, and your patient already requires a blood transfusion. It can be hepatotoxic, and your patient already has abnormal liver function. It can cause arrhythmia and marrow suppression. So I can tell you already that I won't be recommending ribavirin or favipiravir or any other drug still under investigation."

He pursed his lips, shrugged, and nodded his okay.

"If this proves to be a second case of Crimean-Congo virus, we identify that as a nairovirus, a member of the Bunyaviridae order. Its most common vector is *Hyalomma* ticks in Eurasia, but apparently drug users can pass it by sharing needles. The greatest risk of nosocomial infection is from splash exposures and needlestick injuries, so make sure you're using the appropriate precautions. Protect yourself and your staff.

"If and when you come up with a positive antigen, RNA, or antibody test, then we'll notify the CDC. Now I need to review the chart and examine the patient. I'll leave a brief note on the chart, and my longer dictation should be back by morning."

FIFTEEN

Anthropocene Agonistes

First rule of change is controversy. You can't get away from it for the simple reason all issues are controversial. Change means movement, and movement means friction, and friction means heat, and heat means controversy.
—*Saul Alinsky*, Rules for Radicals

The winter daylight was slinking away like a pickpocket as the appointed hour approached for the next gathering of the energy group collaboration. And it was sleeting this time, the sunlight pale and washed-out, the trees haggard and shivering in the northeast wind. Stamping their feet and hunching their shoulders, a pair of zealots by the front door of Science and Engineering held up a wide poster board declaring, "The End Is Nigh." But it turned out to be just a spoof by a couple of undergraduates waiting under the overhang. Ligachev stopped to talk with the sign-holders and laughed out loud, shaking off his umbrella before heading inside to set up.

Jake and Abbey arrived next, rolling their eyes and smiling at the faux protesters, then peeling off their coats as they walked into the main first level of Lehman. Abbey gave Jake a quick peck on the cheek before going up three levels to sit with Brian and Rachel. Jake was glad to see an even larger crowd than last time.

After a few moments' conversation with Mikhail, Jake turned to face the gathered group. "Good evening, and good to see so many people alarmed that the end is nigh—or is it?" he said, turning to Mikhail, who shook his head firmly, *no*.

"Our speaker is again Professor Mikhail Ligachev, who is going to re-assure us that in spite of the daunting issues ahead, solutions abound. The unvarnished reality is simply that we need to educate ourselves and those around us, then walk the talk, both personally and politically. It would be hypocritical not to make changes in our own lives. It would be impolitic not to join together to leverage community and governmental responses as well.

"So, let's get ready for a schooling on climate." He raised his chin to Mikhail and sat down. Mikhail, sporting a green bow tie and a tan suit coat with reinforced elbows, was already busy writing on the board:

The Great Oxidation

"All of you who enjoy breathing should feel indebted to the 'Great Oxidation Event.' About 3.23 billion years ago, cyanobacteria evolved and prospered. They developed photosynthetic capabilities, which entailed taking up carbon dioxide and discharging oxygen as a metabolic product, just like the other algae and plants that subsequently evolved from them.

"This was significant for the land and ocean, because there'd been no appreciable amount of diatomic or molecular oxygen until then, and iron atoms floated freely in the salty ocean without rusting. But oxygen is impetuously reactive, forming compounds generally called 'oxides' with all but four elements, all four of which are noble yet pretty snooty and stand-offish gases."

Jake smiled.

"Immense amounts of iron then rusted once free molecular oxygen reached about 2.5 percent of the atmosphere, forming among other agglomerations the beautiful banded iron formations geologists love so much.

"Oxygen stabilized near its current twenty-one percent level less than a billion years ago, after an overshoot, which may have vigorously augmented wildfires for some protracted interval. Once paired, or diatomic, oxygen developed, then tripled, or triatomic, molecules of oxygen called 'ozone' could be generated, particularly at the mid-level stratosphere. The pivotal significance of this is that this life-preserving ozone filters out ninety-seven percent of the punishing ultraviolet from the Sun, preventing most UV radiation from reaching the surface. This enables the evolution of creatures outside of the protected depths of the ocean and the crevasses of the terrestrial environment." He stopped pacing, tugged at an ear, then reversed direction.

"Just as we've had hothouse stages in our history, we've had icehouse episodes as well, evocatively nicknamed 'Snowball Earths.' These ancient events occurred at least twice, long before the advent of multicellular life. The earliest appearance corresponded with the Great Oxidation Event, with the logic that early photosynthesis pulled a lot of carbon dioxide out of the atmosphere, leading to radically colder temperatures. That episode took place nearly 2.22 billion years ago, again with evidence from sedimentary deposits of glacial origin at tropical paleolatitudes, frictional striations on rock formations from moving ice, random 'dropstones' left behind by melting glaciers or icebergs, and other enigmas not otherwise explicable.

"These icehouse Earths were not for the timid of heart. Either the entire surface of the land and ocean froze, or perhaps there was a slushy belt around the equator. Once this brilliant white planet appeared, most of the solar flux was reflected away by this extensive and intensive albedo. How did our planet escape permanent transformation?" Mikhail looked around the auditorium but saw no one audacious enough to speculate.

"The escape can possibly be attributed to a large asteroid or comet collision or sufficient volcanism breaking through the ice to significantly escalate carbon dioxide levels. Absent some sort of intervention, our planet could have become permanently frozen—comparable to the moon Europa, which orbits Jupiter—with salt water beneath ice, albeit perhaps teeming with life.

"The staggering extremes of sea level are my primary focus here. During the time of an icehouse Earth, the seas were down some 1,700 feet, while fifty-six million years ago during a hothouse Earth, the sea level rose two hundred feet above the current level." Again, Mikhail hesitated, apparently assessing the room.

"Why do I focus on all of this ancient history? Because we live in an uncaring universe, on a planet capable of violent extremes. Because we've been lulled into believing the Earth will protect us, come what may, the ludicrous folly of the discredited Gaia hypothesis. Because we too casually assume life itself cannot nudge the climate in uncomfortable directions."

If cyanobacteria can do it, so can we, Jake thought. *And Abbey and I are going to bring a child into this world.*

"Because we aren't willing to accept responsibility for being stewards of the planet and all the life on it, if only because we wish to survive. We only survive if almost all other life survives." The crowd stilled as Mikhail stopped pacing for a moment, shaking his head.

Then he scrawled on the board:

Five Mass Extinctions

"But, sadly, diversity of life does not ensure survival. Since the Cambrian over half a billion years ago, five mass extinctions have occurred, defined as the loss of more than half of all extant species. I will describe only two of these events.

"The third in the sequence was the worst, called the Permian-Triassic or simply End-Permian extinction, or the 'Great Dying,' and took place 252 million years ago. This was attributed to mass volcanic carbon release in the forms of carbon dioxide and methane, with high temperatures and profound oceanic acidification. A full ninety-six percent of marine species and seventy percent of terrestrial species were marooned forever in the fossil record." His face appeared serious, his facial folds deepening.

"The last of the five extinctions is the Cretaceous-Paleogene or K-Pg extinction sixty-six million years ago. That was associated with a greater than seventy-five percent loss of dinosaurs on land, plesiosaurs in the ocean, and pterosaurs like the pterodactyl from the skies. My favorite fossils from this era are the innumerable ammonites, and as a matter of fact, I have several specimens in my office. The mammals and birds came through admirably well and emerged as the dominant land vertebrates. Vertebrates, of course, are animals with an internal or endoskeleton, especially a backbone."

Which some politicians apparently lack, Jake thought.

"As a cautionary note, we're now releasing greenhouse gases at a rate equaling or exceeding that of the major extinction events not involving an asteroid. This led the IPCC in 2014 to make this observation: 'While only a few recent species extinctions have been attributed as yet to climate change (high confidence), natural global climate change at rates slower than current anthropogenic climate change caused significant ecosystem shifts and species extinctions during the past millions of years (high confidence).'

"That opinion has not improved in later iterations of their reports. Contemporaneously, a sixth extinction event may be underway, as cited in a book with that very title by a science journalist named Elizabeth Kolbert. She states that a 'third of freshwater mollusks, a similar fraction of sharks and rays—more than a quarter of all mammals—a fifth of all reptiles and a like proportion of birds are headed towards oblivion.'

"We should not be overconfident by reason of being in the primate order, as many of our fellow primates and great apes are also at risk of failing. Orangutans and gorillas are in this category. Another term for the widespread loss of the world's wild fauna is 'biological annihilation'—an inadequate breeding population—which necessarily occurs before absolute extinction.

"Since 1970, Earth's various populations of wild land vertebrates are estimated to have lost, on average, sixty percent of their numbers. The consequential point is the current rate of accelerated species extinction is already proceeding faster than it has in some prior mass events, raising the question of whether or not we're actually committing the world to a sixth annihilation event." Mikhail leaned over to write lower down on the whiteboard:

Tree Species Extinctions

"We have an obvious concern about megafauna—the Earth's large mammals—because, well, we fit that definition. But terrestrial plants make up the majority of the world's biomass and contain five billion metric tons of carbon, which is four times more than all the other organisms put together.

"In 1987, UK-based Botanic Gardens Conservation International was founded. The organization works with eight hundred botanical gardens in 118 countries. In 2017, they published a global list of more than sixty thousand tree species from data supplied by member organizations, with the intent of identifying species at risk and organizing efforts to preserve them.

"More than half of all types of trees occur in a single country. A conservation status has been verified for only about a third of them. The working consensus is that fifteen percent of these plant species are at risk. These trees face threats from either overharvesting for furniture and construction, or land clearance coupled with their use as a source of subsistence fuel." Mikhail turned and scribbled another subheading on the board:

The Holocene Epoch

"We are clearly up to the 11,700-years-ago dawning of the Holocene we are living in now, which may be the most stable climate interval of the last 650,000 years. This slide summarizes the surprising timeline for how we've been affecting climate for much longer than all of you might have thought."

Holocene Epoch Timeline

- ~10,000 yrs. ago onset warming in interglacial MIS-1 with deforestation + CO_2
- ~5,000 yrs. ago further warming out-forcing orbital cycles, with rice farming + CH_4
- ~200 yrs. ago Industrial Revolution with further 1.15° C heating by 2023
- 2009 Arctic lightning up 20-fold, some first tundra + taiga fires

Fabled NW & NE Passages Either Side of Greenland

- 2008 opened up first time to free summer navigation
- discounting 1906 Amundsen precedent [ship spent winter in ice]
- NW passage saves 2,000 miles + transit Panama Canal
- NE passage saves 2,000 miles + transit Suez Canal

Last surviving hominin thriving with predictable agriculture + hydrologic cycle

"About ten thousand years ago, we began leveling forests for the purposes of fuel and construction wood, and for slash-and-burn agriculture, all releasing carbon dioxide. Five thousand years ago, we ratcheted up warming—out-forcing the effect of orbital cycles—with rice farming and its resultant methane release. The combined effects of the two trace gases of carbon dioxide and methane in the atmosphere boosted worldwide temperature an estimated 0.5 degrees Celsius even before the onset of industrial society."

"Mikhail, could you explain what MIS-1 is?" Jake asked, knowing this was an important clarification for the audience to hear.

"Glad you asked that question, because I need to talk about stable isotopic dating next. But the answer you're looking for is Marine Isotope Stage-1—essentially the Holocene epoch or our current interglacial period—because the shells of mollusks contain useful and informative oxygen isotopes. These marine stages are numbered in reverse chronological order, which means the prior glacials and interglacials have higher numbers, often including subsets within each, not to put too fine a point on it."

Rachel expressed her curiosity. "And what exactly was it that Amundsen accomplished?"

"Norwegian Roald Amundsen and his crew set out on a voyage over northern Canada by way of the Northwest Passage, accepting the fact that they'd have to spend one whole winter with their wooden ship stuck in the ice near the halfway point, which is not what most people would voluntarily choose.

"Historians debate the exact onset of the Industrial Revolution, but during the last two centuries preceding 2023, the planet's average surface temperature climbed another 1.15 degrees Celsius, which brings us to our current plight." He turned again to write.

Stable Oxygen Isotopes

"Scientists concentrate on getting timelines right, employing a multitude of overlapping approaches. You've no doubt heard of 'heavy water,' which involves the deuterium isotope of one of the two hydrogens in a water molecule. Deuterium is simply a hydrogen atom with an extra neutron, which increases the weight of the whole water molecule by that amount. But there's actually a second way to make a water molecule heftier, as summarized in these slides."

Delta-Oxygen-18 Dating

- O-16 [oxygen-sixteen] has 8 neutrons, water evaporates easily [lighter]
- O-18 [oxygen-eighteen] has 10 neutrons, water precipitates easily [heavier]
- ratio of O-18 to O-16 called $\delta^{18}O$ [delta-oxygen-18]
- O-18 + O-16 assayed in marine sediments, ice cores, fossils

During Glacial Intervals, Cooler Temps Extend Toward Equator

- O-18 rains out earlier
- higher proportions of O-16 sequestered in northern ice, sediments, fossils

Exact oxygen ratios in sediments + shells + fossils correlate with global extent of paleowater trapped in ancient ice sheets

---

"It looks complicated, but really it's not. Oxygen-18 has two more neutrons than oxygen-16. So if it's incorporated into water, this heavier molecule finds it harder to evaporate, and is also more eager to precipitate out of a cloud. When a clam uses this type of water to synthesize calcium carbonate and ends up with these oxygen atoms in its shell, those atoms can persist not just during the clam's lifetime, but in its fossil for millions of years.

"So on one hand," Mikhail said as he turned his right hand palm up, "when the climate regimen is colder, as it is during a glacial period, the oxygen-18 water rains out at even lower-than-usual latitudes of the globe and altitudes in mountains. As a result, the oxygen-16 water approaching the poles reaches a higher sequestered proportion. The ratio of oxygen-18 to oxygen-16 is what's important. This is called the 'delta-oxygen-18.' It functions as a 'paleothermometer,' recording how cold the planet was when that particular rain or snow precipitated and ended up in glaciers and ice sheets.

"On the other hand," Mikhail said, turning his left hand over, "when the climate regimen is hotter, as it is during an interglacial period, the water molecules with incorporated oxygen-18 evaporate more easily. They may get much farther north and higher in elevation as they take part in the hydrologic cycle. Thus, these molecules with added oxygen-18 leave an ancient record of themselves." He scratched his nose.

"I should emphasize that these isotopes are stable, not radioactive. So there is no decay and this measurement is in no way comparable to radio-carbon dating, which is a horse of a different color. Perhaps, a palomino." Then Mikhail wrote on the board again:

Other Paleothermometers

"The existence of multiple lines of corroborating proxy temperature data is reassuring to scientists. One of these corroborators, termed 'paly-nology,' uses data from ancient pollen and spores."

Jake reminded himself to transcribe these important notes he'd been taking taking onto his computer within a day. From sad prior experience, he knew if he waited too long, he might not be able to resurrect the logic of his shorthand.

Mikhail continued. "Conifers, technically 'gymnosperms,' rely on un-assisted pollination and therefore release prodigious amounts of pollen.

They predominate in cooler areas and eras. Deciduous trees, technically 'angiosperms,' have coadapted with pollinators like insects and birds, and therefore can get away with releasing much lesser quantities of pollen. These tend to thrive in warmer areas. In lake sediments or in bubbles of glacial ice, the ratio of conifer to deciduous pollen represents another way of estimating ancient temperatures.

"Likewise, the presence of dust attests to either drought or windy periods.

"Sulfates from a volcanic eruption correlate with independent dating from a geographically remote volcano by using its unique radioisotopic signature.

"Ash from wildfires is an indicator of widespread conflagration.

"The field of 'dendrochronology' is the study of tree rings, which have alternating growth and dormancy bands. Overlapping tree records of different vintages often can leapfrog and generate a year-by-year record of drought and wildfire going back centuries.

"Note the typical consensus between all of these independent lines of evidence. This offers incontrovertible proof when dealing with those who question climate change," he said, adding a performative sigh.

"'Speleothems' are those beautiful structures in caves—stalactites and stalagmites and other exotic features. When cored, their simple geometry makes them particularly useful for dating with uranium–thorium ratios, delta-oxygen-18 as just described, and a parallel technique with stable carbon-12 and carbon-13 isotopes. And the slow percolation of water through caves halts as the water reaches stasis during either prolonged drought or with frozen ground above." Turning, he wrote:

Marine Deoxygenation

"But first, we should tackle several more oceanic puzzles. Apparently, we've decided that it's not enough to warm, raise, and acidify the ocean; we're also detecting climate-change-driven oxygen loss, or hypoxia, in certain, more vulnerable areas. Reckoning is that this will likely be widespread in coastal areas by 2030 or 2040. Some areas of the ocean may become all but uninhabitable for certain species. Marine mammals like dolphins and whales breathe by surfacing, but fish and crabs rely on dissolved oxygen coming from both terrestrial plants and marine algae, mixed by waves and wind."

Jake shook his head in silent admiration for this prof's smooth, polished delivery—despite the anxiety-inducing content.

"First, a warmer ocean holds less oxygen, especially in the summer, as oxygen, nitrogen, or any molecule absorbed into the water will have what is called a lower 'partial pressure.' Second, as water heats up, it expands, and the resultant layering, or stratification, of warmer water on top of colder water may resist circulation. A third factor is that any oxygen-breathers in warmer waters will increase their rate of metabolic activity and thus their rate of respiration, or oxygen use. Finally, increased nutrient-loading of nitrogen, phosphorus, or organic matter from farming, fertilizer, wastewater, or other runoff may lead to an algal bloom. As the algae die, aerobic bacteria consume them, further depleting oxygen.

"All four of these factors drive down the dissolved oxygen that's so crucial for ocean-dwellers." Mikhail set the remote on the lectern in order to remove his suit coat, which he set on a chair.

Jake thought to himself, *So, temperature, stratification, metabolism, algae.*

"Marine hypoxia is defined as any level less than two milligrams of oxygen per liter. On a global basis, the ocean loses a billion tons of oxygen annually. For example, serious levels of oxygen deprivation are anticipated for the Pacific Northwest in 2035. We need to be concerned as a civilization—not just a country—because combining oceanic warming, acidification, and deoxygenation creates conditions equivalent to the End-Permian mass extinction 252 million years ago. Recall how that went for life on Earth, most devastatingly for marine creatures." Again Mikhail used his quick, athletic scrawl:

Marine Cacophony

"Okay, time to talk about noise pollution. That's right, stop all that whispering up in the top row." Jake turned around to see Brian and Rachel looking sheepish. "All jesting aside, anthropogenic marine cacophony is creating a 'sonic sea' damaging to many kinds of animals. In 2019, the federal government decided to allow offshore drilling on the Eastern Seaboard for the first time in three decades. Five companies received permits for seismic mapping, but the courts fortunately blocked their efforts.

"Such a ship deploys anywhere from a dozen to forty-eight air guns, triggering repeated blasts of pressurized air, a reverberating sound pushed miles deep into the ocean floor. The echoes rebound to a separate array of hydrophones for recording. These air guns may 'fire . . . every ten seconds around the clock for months at a time' as quoted by Douglas Nowacek, professor of marine conservation technology at Duke. On this slide are listed the agonal intensities and biological effects of all this racket on sea life."

Marine Cacophony
Comparative Intensities in Decibels, or dB

- seismic blasting estimated up to **260 dB**
- container ships max at **190 dB**
- space shuttle launch reaches **~160 dB**
- every 10 dB increment is 10 times louder

Effects Air Guns + Ship Sonar

- impaired hearing, brain hemorrhaging
- whales, dolphins, fish, squid, octopuses
- drowning out communication sounds
- tiniest, even plankton, including krill, affected

Max Baerbock spoke up in his baritone. "In the summer of 1969, I was working at a summer camp in central Florida, and we took busloads of kids to watch Apollo 11 take off. We were about two and a half miles from the launch site, and I still had to cover my ears. But I see you have the space shuttle launches listed at only 160 decibels," Max said, and shook his head.

Mikhail nodded somberly. "Perhaps the space shuttle is louder than the older Apollo rockets. But still, seismic blasting remains an inconceivable hundred decibels louder. A 2017 study—I don't recall the reference offhand so I'd have to get back to you—used a noise stimulus quieter than an air gun. The study still found that the noise killed nearly two-thirds of zooplankton within a three-quarter-mile radius. And zooplankton are the prey for species ranging from tiny shrimp to great whales.

"Naval sonar is also problematic. This technology makes use of the fact that sound used both for communications and sonar travels five times faster and farther in water than in air. The US Navy uses a submarine and mine detection system known as Low Frequency Active Sonar, or LFA, throughout eighty percent of the world's seas. It relies on a vertical array of eighteen emitters to pulse low-frequency blasts of about 215 decibels roughly every minute. Many whales have tried to escape all of this noise by heading for the Arctic and Antarctic, but sadly, these great oceans no longer supply the refuge they once did. Polar ice is melting, while seismic exploration and ship traffic are increasing."

For reasons known only to Mikhail, he went on to post his next term in red:

Marine Plastification

"Plastification—a real word—is another assault on the ocean, just as it is on land. More than ninety percent of plastic is still not appropriately— or able to be—recycled. And corporations have generally misled the public about the recyclability of plastics in order to continue to synthesize and sell them to consumers. At the same time, other countries outside the US are rebelling against accepting our discarded trash."

No surprise there, Jake thought.

"And individuals still engage in 'aspirational' recycling with material that just has to be separated out and discarded, gumming up the whole process.

"An estimated 120 megatonnes, or million metric tons, of plastic already reside in the ocean. The visible plastic accumulates mostly in large, subtropical gyres. But what we see is deceptive, because only about one percent of the plastic lies on the surface. All of this macroplastic poses a serious risk for marine mammals and turtles, who may become entrapped, suffocated, or starved from obstruction of the gut.

"Plastic microbeads are those multicolored spherical balls only a fraction of a millimeter in diameter, quite popular for a while in exfoliating facial scrubs, cosmetics, soaps, and even toothpaste. They're problematic, since they're small enough to pass through filters at wastewater treatment plants, have a troublesome tendency to absorb and concentrate pollutants, and look enough like fish eggs to get eaten by a variety of aquatic organisms.

"In 2017, the UN calculated that by midcentury, plastic will outweigh all the finfish in the ocean. Unfortunately, we're still on track to hit that target. With ninety percent of oceanic stocks depleted from overfishing, plastics are accumulating while fish are dwindling." The whole room grew silent in response to this news.

"Okay, time for a break. When you come back, we'll talk about the Sun, even though it's a rainy day today. Please be back in ten minutes, as I pledge to start on time."

SIXTEEN

Velvet Glove

A nation can survive its fools, and even the ambitious. But it cannot survive treason from within. An enemy at the gates is less formidable, for he is known and carries his banner openly. But the traitor moves amongst those within the gate freely, his sly whispers rustling through all the alleys, heard in the very halls of government itself.
—*Marcus Tullius Cicero*

Restless as a long-tailed cat in a room full of rocking chairs, the sunglasses guy paced in circles around the attic, making fists, backing out of rooms for no good reason, opening drawers, not knowing what he was looking for. The air was heavy and smelled of mold as dribbles of sweat from his armpits ran down his ribcage. He threw a treat to his dog, who snapped it up in midair like an alligator.

The animal was no better: panting, whining, twisting to follow his every veering. Too hot, yet not dark enough to go out for a run, the release he craved—they both craved. He gradually calmed himself and sat his elbows at the computer terminal desk, fingers splayed on each side of his forehead. He entered separate passwords for each of the three screens and got to work.

He flicked a piece of dust from the screen on the left. Unsatisfied, he pulled out the soft cloth for cleaning his glasses and thoroughly cleaned the entire monitor.

It was all a question of timing. Disrupting a nuke operating at full power was his obsession, but context was critical.

A strike at the moment electricity demand peaked would amplify the impact.

A snowstorm or flooding would accentuate vulnerability.

A radiation release would panic millions, wafted to the city by winds from the northeast.

A convocation of world leaders in the capital would be the veritable jackpot.

He pulled up his files of hacked phones, poring over the software-transcribed conversations of his targeted players. Some he had followed for so many months that he felt the beginning of a stalking relationship with them. A select few had received nicknames. He'd set them up on an organizational chart. Those identified as administrators were of lesser interest, but senior supervisors and reactor operators were a different kettle of fish altogether.

Delusion and confusion were his organizing principles. When he set his plan in motion, success would depend on duping many of the monitoring circuits into giving innocuous status readouts. And he would have to simultaneously corrupt many of the communication circuits within and extending outside the plant—most importantly, the cell phones of the seven key players he'd identified.

Backup diesel generators. Heat exchangers outside the reactor containment. Automatic controls for the cooling water. Communication circuitry. Humans prone to error.

So many attractive targets.

SEVENTEEN

Planetary Forcings

Outside, the sleet continued to fall. People had gathered in the hallway or just outside the entrance or even made their way back inside prematurely, still chattering about noise and plastic pollution. They also talked about the basketball game tomorrow night against Virginia State.

Jake called the room to order after the short recess, and Mikhail set right back to work with his sleeves rolled up.

"One of the common queries I hear from people is whether or not variations in the Sun's intensity could explain what's going on. Climate denialists often attempt to delude people with this sort of propaganda, frankly. Solar radiation, of course, varies in intensity over a decade or so, and in spectral distribution over thousands of years. Since the 1970s, we've been observing the Sun through the satellite era and getting accurate and voluminous data." Erasing again, Mikhail put up:

Solar Variation

"Based on the solar activity forecast, heightened sunspot activity was expected in about 2025, but it arrived instead in 2024, the strongest in twenty years, with knock-on effects on satellites and electric grids. On average, the main sunspot cycle peaks every eleven years. The last two were in 2001, followed by the last one in 2014, which was relatively anemic.

"Paradoxically, more sunspots indicate an accentuated solar flux. Sounds counterintuitive, but while the spots are visibly dark, the surrounding areas of the Sun's surface are putting out power that more than compensates for the lesser contribution from the spots themselves.

"But the total solar variance is equivalent to only 0.1 percent, or about 1.3 watts per square meter, at the top of the atmosphere, peak-to-trough, or maximum-to-minimum. During recent down cycles, the Earth's temperature kept rising. The second nail in the coffin for us is that if the up cycle of the Sun's intensity was responsible for warming, then the upper atmosphere would increase in temperature as much or more than the lower atmosphere, which is simply not the case.

"So any recent changes in solar brightness have been too weak and in the wrong direction to explain climate change. Case closed." Mikhail scrawled:

Land Clearance

"United Nations data support the idea that a quarter of our land area has been degraded by overgrazing, excessive plowing, and rapacious timbering, and that the combined deforestation and desertification result in about eleven percent of ongoing greenhouse gas release. Trees and their root systems and their associated mycorrhizal fungi are unhappily being converted from carbon sinks into carbon sources.

"Professor Higgenbotham tells me she previously discussed the contribution of cement production to GHG emissions. And there's no question we could improve our nutrition and diminish our carbon footprints by modifying our meat-intensive diets."

Then Mikhail erased the prior heading and wrote:

Human Overpopulation

"But now, I want to shift focus to the issue of population growth, which many environmentalists have shied away from in the last few decades. In 2022, the world reached eight billion people. But I was born in 1950, when global population was only 2.5 billion, which means the population has more than tripled in my lifetime.

"At the same time, more than eighty countries either remain static or are below the population replacement rate, including ours. 'Replacement-rate' fertility is conventionally set at 2.1 children per couple, since not everybody chooses to reproduce or survives long enough to have kids. But in 2020, the American fertility rate was 1.8 per woman, while China's was 1.6, perhaps with contribution from the infamous coronavirus pandemic.

"The solutions for overpopulation are pretty straightforward: namely, education in literacy and numeracy for girls and women, plus providing the whole spectrum of reproductive health care to both women and men. An important piece of the puzzle is the concept of long-acting reversible contraception, or LARC."

Mikhail continued. "Some of the IUDs can stay in place for a decade. And the single small subdermal progestin implants last for at least three years, with some data showing efficacy for five. Both display remarkably low pregnancy rates, much less than one percent per year."

Jake thought back to when he and Abbey had concurred it was time for her to remove her implant.

"And here is a fine example of why it's not sufficient to just act personally. The infamous 1973 Helms Amendment to the Foreign Assistance Act prohibits any medical aid to a foreign country that includes provision of or even provides information about abortion services. Therefore, it's a political imperative that our foreign policy legislatively support all information and services comprising reproductive health care, including abortion services in those countries and cultures that allow the procedure. To this end, we need to do away with the Helms Amendment, and that will only happen with lobbying and all-out support for candidates with whom you see eye to eye."

He sighed. "Of course, we need to reverse our country's 2022 Supreme Court decision and restore abortion rights by getting them codified into law.

"This is an ethical issue and a medical issue, but fundamentally not a religious controversy, because none of the foundational texts of major religions outlaw abortion. And, in polling, most Americans feel that women and couples should have access to pregnancy termination up to a gestational stage supported by a medical and societal consensus. Furthermore, vasectomy and tubal ligation are low-complication procedures which should be freely available to any competent adult."

A few in the audience frowned or looked pensive at these remarks, but no one voiced complaints. Rather, Jake heard a small smattering of applause. Several men shifted uncomfortably in their seats.

"Okay, let's get back to the science of climate." Mikhail turned and scribbled:

Permafrost

"I like to think of permafrost as our *underground* planetary icebox. It occupies about a quarter of the Northern Hemispheric land surface, and in some areas, is frozen as deeply as six hundred meters or nearly two thousand feet. A shallow, 'active layer' a vertical foot or two at the surface might soften in the summer, but refreeze each winter. This layer will nonetheless allow seasonal growth of shallow, horizontally rooted vegetation.

"Permafrost contains about twice the organic carbon encompassing the entire atmosphere. So mammoth positive feedback would be the consequence of its melting; that is, the release of carbon would hasten its further melting. A thaw converts a long-term carbon *sink* to an active *source* of carbon dioxide and methane. The melting of permafrost would triple the amount of carbon in the atmosphere."

"If the surface of the ice melts significantly, then small ponds become part of the so-called 'thermokarst' landscape and serve as lenses channeling sunlight into the much deeper subsurface of the permafrost. Coastlines erode. Ground liquefies. As a consequence, buildings and roads collapse. 'Drunken forests' result, when groves of trees lean over because their roots have become destabilized.

Jake smiled at the idea of drunken trees and drew a quick sketch.

"A less well-recognized consequence of melting permafrost would be the release of an estimated fifteen million gallons of mercury sequestered in ice. Imagine fifty Olympic-sized swimming pools containing this toxic heavy metal. This represents roughly twice the mercury in other soil types, the ocean, and atmosphere combined.

"Nitrogen and phosphorus releases from melting permafrost are also significant. Their admixture into fresh or marine water intensifies pollution, fostering harmful algal blooms.

"Another interesting aside involves permafrost in the upper crust on Mars. But at least there, it'll remain stable. However, as climate journalist Brian Kahn observed about our terrestrial predicament: 'Permafrost is becoming, well, less permanent.'"

Mikhail wrote again on the board:

Methane Clathrates

"Fewer people have expressed short-term concerns about a positive climate feedback signal from marine methane clathrates, which are also called 'methane hydrates' or 'ice that burns.' But I like to think of this unusual methane sink as our *underwater* planetary icebox. You may have seen samples of it in a lab that look like ice but can be lit on fire. The amount of carbon trapped in the lattice structure of clathrates at least matches the amount of carbon already in the atmosphere. How do clathrates form, you may ask?" Mikhail looked around, evidently seeing interest.

Jake was confident the audience was right there with him.

"Imagine gaseous methane at, say, one hundred degrees Celsius, migrating up through subsea geological faults and encountering cold, pressurized seawater. A single methane molecule may be trapped in a crystalline 'bird cage' or lattice of water ice.

"If released, that captive methane molecule will capture eighty-six times more heat than a single molecule of carbon dioxide over its first several decades, twenty times as much if it persists for a century. At least methane remains for a much shorter time in the atmosphere, far shorter than that of CO_2.

"The third most important heat-trapping gas, nitrous oxide, will trap about 280 times more heat than carbon dioxide on a molecule-to-molecule basis. There must be a conspiracy of dentists trying to squelch this information. Instead of venting it from the office, I wonder if we might capture it for the purpose of disposal or deconstruction. These three gases are often aggregated as 'CO$_2$ equivalents' for the sake of calculations."

Mikhail again wielded his marker:

Blackbodies in Physics

"Moving on then, what in the world is a blackbody?" He turned and drew a circle on the board with a rapidly vibrating arrow entering on the left and a more relaxed, squiggly arrow exiting the opposite side. "And what is blackbody radiation?

"Imagine a bowling ball sitting on a table. To be precise, this is an incomplete blackbody, because you can see it. This means it's reflecting a modest amount of visible light. A perfect blackbody would be invisible, reflecting nothing and absorbing all the radiation of the entire electromagnetic spectrum. But it's a bowling ball, after all, so it doesn't generate any light of its own. If you focus a bright light or laser on a blackbody from a distance, its temperature will begin to rise. And we know its temperature could not start at absolute zero, since the 'zeroth' law of thermodynamics states that you can't get to that starting point even if you wanted to."

Brian whispered to Rachel, "Why does he even have a bowling ball on the table in his living room?" and she had to suppress a giggle.

"Eventually, our incomplete blackbody will get warm enough to stabilize its temperature. The light or laser is warmer than the ball, but the temperature does not continue to rise. Why should that be?" He raised both hands to indicate that the question was not rhetorical.

Jake paused for a moment, then suggested, "As its temperature rises, it would begin to give off heat by conduction into the table, convection into the air flowing past it, and by infrared radiation to the surrounding room."

Mikhail crossed his arms. "In fact, was it not already transferring heat by all three of these mechanisms—even before the light or laser was turned on? How do you explain that?"

Jake responded, shrugging. "It's shedding heat by the same three maneuvers, just at a higher rate."

"Excellent, yes, heating up until a new equilibrium is achieved, with energy inflow balanced out by energy outflow." Mikhail stood for a moment at the board, probably wondering if he should post another title, but apparently deciding not to.

"So now we've supported an intuition that any object, any sample of matter, can both absorb and emit electromagnetic energy such as ultraviolet, which causes tanning and sunburn. Also visible light, which led to the evolution of multiple types of eyes. And infrared, commonly known as 'heat radiation.' The ideal blackbody would reflect none of these types of energy but could heat up enough to emit all of the same classes of energy. But almost all blackbodies are incomplete, less than ideal. Nonetheless, they undergo similar exchanges.

"The blackbody then heats up from this flux, or flow, of energy, and begins to emit more energy in turn, trending toward an equilibrium. The lowest energy is emitted because, well, this is a bowling ball, right? Not very smart and in fact lazy if no bowling pins are around for it to knock down. So it emits infrared energy in all directions, even though it receives energy from only one point of the compass. Clear, so far?"

Rachel could not hold back. "But if you got it hot enough, it could give off radiation in the visible part of the spectrum, even up into the ultraviolet and higher, yes?"

Mikhail nodded and smiled in agreement. "Inarguably so. Now, let's drop one specification but add some others. Just a regular bowling ball is now acceptable, even sporting some kind of white logo, but its location is now in orbit out in space around the Sun. Our bowling ball is rotating on its own axis, so when the logo is facing the Sun, the albedo of the bowling ball is greater—so less energy is absorbed. Of course, there's a pretty hard vacuum in space. And in the shade behind the bowling ball, like the dark side of the Moon, the temperature is only three kelvin above absolute zero. Pretty chilly." Mikhail mimed shivering. "The Sun is bombarding it with a nearly full spectrum of energy, so now . . .," he said, pointing to Rachel, "how is the bowling ball going to achieve thermal equilibrium?"

"Well," she responded slowly, "conduction is out, assuming the table has gone missing. And convection is impossible without measurable gas around it. So it can only reach a stable point by giving off more and more infrared."

"So where do you think that I am going with all of this?" He gave the class a chance to respond. "Anybody?"

Brian couldn't resist. "Bowling?" he said, which garnered laughter even from Mikhail, but also a punch in the arm from Rachel, along with a whispered, "*Du Idiot.*"

"It might be safer to answer my own question. So let me now announce my own conception, the Thermos Bottle Theory of Global Warming, of which I am proud, but humbly so. Earth is a ship in a bottle, except that our world is inside a transparent thermos bottle. The average temperature of the surface of this rocky planet is 59 degrees Fahrenheit or 288 kelvin. The top of the atmosphere, or TOA, is the inner wall of the thermos bottle, the distant universe is the outside wall, and in between the two walls exists a cold, hard vacuum. The temperature of a vacuum is irrelevant, but the temperature of any relatively close targets is pertinent if you are trying to emit infrared, since they may be radiating right back at you." He held up his hands as if to block that infrared light.

"Fortunately, the planet may freely radiate infrared to its heart's content, even visible light, such as that of wildfires, methane flaring, bioluminescence, volcanic eruption, lightning, and so forth. And remember that the surface of the planet immediately reflects a full thirty percent of the incoming solar. Or the clouds reflect it, much like the bowling ball logo would.

"The spectrum of sunlight arriving at the TOA includes five percent ultraviolet—forty-six percent visible and forty-nine percent infrared—plus some other miscellanea. Note the vision we've evolved as humans is the part of the spectrum between 390 and 700 nanometers, which I will remind you are billionths of a meter. The evolution of vision followed Willie Sutton's Law on at least twenty separate occasions. Insects and some birds of prey may see into the ultraviolet and pit vipers can sense infrared, but most sorts of eyes are centered in the visible portion of the spectrum reaching the planetary surface."

A question came from someone in the back. "Who was Willie Sutton?"

"Mr. Sutton was a bank robber," he said, and guffawed. "After he was finally captured, a group of reporters approached him in his jail cell and allegedly asked why he robbed banks. 'Because that's where the money is,' he replied. It just makes sense that eyes would evolve to use the more intense part of the available spectrum. Willie Sutton's Law. Except later, he complained he'd never really said that. It was just some reporter's trick.

"But back to the Earth achieving thermal equilibrium. For all the random sources of emitted visible light—the wildfires and lightning I mentioned—the overwhelming bulk of energy is given off as infrared. This long-wave radiation is attempting to escape from a blackbody—an incomplete blackbody—and the heat-trapping gases just get in the way, like a blanket covering a window.

"Each of the special molecules we are discussing, like carbon dioxide, methane, nitrous oxide, and all the rest, absorb the photon of infrared in an outer electron orbital and then quickly emit it in any random direction. So basically, half the time infrared is emitted back down toward the planet, which is creating this problem we call global warming.

"Here's a question for you. The Sun is a much hotter blackbody with a temperature of 5,770 kelvin. So, why are we not fried like an egg in a frying pan?" He smiled to himself but received no attempts at an answer.

"Because Earth absorbs energy as a rotating opaque sphere from only one direction, but radiates energy in all directions, a four-to-one ratio. Because the net albedo of the world is about thirty percent, and successfully reflected radiation does not heat up the planet at all."

Mikhail slowed his speech abruptly. "But of the seventy percent of energy which gets absorbed, all of it must get reemitted out past the TOA, or the temperature of the planet will be forced to change.

"The crux of the problem is that if we keep trapping more infrared, the planet will heat up until we reach a new, higher equilibrium. Recall

Venus trapping so much heat through a runaway greenhouse effect that the element lead now exists as a liquid on its surface."

We won't get that hot of course, but jeez, Jake mused, *all the world's ice has melted in the remote past . . . more than once.*

"Finally, a brief commentary on the Anthropocene epoch supplanting the Holocene epoch. This sort of change in nomenclature is adjudicated by the International Commission on Stratigraphy. When biologic and geologic history reaches an inflection point, it's time to move on. The geologic record has laid down recent markers—such as plastics or radioisotopes from above-ground nuclear testing—that future geologists will be able to use in dating.

"And it's clear our species has already modified most of the planet's surface, in some ways irretrievably. So an Anthropocene epoch makes sense to me. And it'll be fascinating to see how all of this plays out.

"Now I think we shall finish up after the holidays. I look forward to discussing clouds, aerosols, the snow and ice of the cryosphere, Mount Pinatubo, the Greenland and Antarctic ice sheets, the saltwater temperature at the grounding line under the Thwaites Glacier, and the Eemian epoch. In short, why we as a species cannot continue with business as usual—with blinders on.

"Have a good break from school. See you here next time in this auditorium—or so I've been told." Mikhail looked at Jake for confirmation, and he nodded his assent.

Outside, the rain had stopped. The usual crew—Jake and Abbey, plus Brian and Rachel with their arms around each other—conferred for a minute.

"On the whole," Rachel said, "people seemed less shell-shocked this time, but perhaps more confused."

"Sometimes I ask a question just to help clarify for others what Mikhail is trying to communicate," Jake pointed out. His friends appeared to agree with him.

Brian spread his hands widely. "Well, if you ask me, I think we should actually go bowling sometime."

The other three laughed.

"No, wait, I'm serious. It's been years for me, so it'd be fun. Really. We've got the whole winter break, so we can tell Professor Ligachev he actually motivated us to independently study the physics of the bowling ball. He'll be impressed," Brian argued.

"Bowling, honestly?" Jake said.

All four friends would remember the lecture and Brian's witty remarks as the impetus for their bowling night. A couple of them only scored in double digits, but they all thought it was hilarious.

EIGHTEEN

The Body Electric

Men may or may not be better drivers than women, but they seem to die
more often trying to prove that they are.
—*Tom Vanderbilt,* Traffic: Why We Drive the Way We Do
(And What It Says About Us)

The poinsettias thirsted for water more often, seemed to know it was last call, and told themselves they deserved one last, long guzzle before being unceremoniously bundled outside on the frozen compost heap. Sharing the counter close to the poinsettias was a crock pot that had slaved away all night on simmer for the sake of baked beans. Jake and Abbey were both up earlier than usual, as their first-to-arrive guest last night had requested a breakfast at dawn, if possible, to coincide with his group's first day of their college reunion. Jake was wholly in accord, an early riser, anyway.

Abbey busied herself chopping up cucumbers and grape tomatoes and finely slicing a single red onion, then mixing in a half cup of balsamic vinaigrette with an old wooden spoon. The lingering scent of the salad was sweet and sharp. "I'm so glad I'm long past the morning sickness so I can enjoy meal prep and eating again." She beamed at Jake.

"I'm with you, babe, believe me," Jake said. *And now she's really frisky in bed again.* "We've got oatmeal or sourdough bread for toasting if anybody wants anything else." Jake paused. "I just remembered that I never heard back from Emmanuelle after I sent her an email about the sunglasses guy. I'd better follow up."

Jake set three places around the end of the main table in the dining room; a cream-colored runner featuring a design of snowflakes in various sizes and colors stretched down its length. Each setting received silverware and soft burlap napkins, plates and bowls, and small glasses. He located a pitcher of pomegranate juice on one side, and salt and pepper grinders on the other. A small bowl of walnuts and a nutcracker flanked by two candles were placed in the center, just for decoration.

Pretty basic setup, he thought, *but more than adequate.*

He went back to the kitchen to get the crock of butter.

"Signs of life," Abbey said in response to sounds of people stirring above followed by the creak of someone descending the wooden stairs at a brisk pace. "Good morning, Mr. MacWhorter," she said to John, whom they had both met last night. "The coffee mugs are here on the kitchen table, together with cream and sugar should you be so bold."

"And good morning to you, guys," he responded. John was a big-knuckled guy, with a narrow white scar on the left side of his forehead and shaggy, dark blond hair. The guy was hairy enough that some poked out of his shirt collar. He was wearing faded jeans, a bright white, long-sleeved shirt with the sleeves rolled up, and cowboy boots. Abbey noticed he wore no rings or any other jewelry, though she also knew his girlfriend was coming up today and would join him for the next couple of nights.

Jake said, "Little bit of snow's coming down out there, nothing that's going to slow anybody down."

"That's good," said John with a glance out the window. "I should still check the forecast because Apryl texted me to let me know she'll be in about noon today. We plan to rendezvous at the reunion site on campus at George Mason, so she'll go straight there first and not stop here." He sat at the table and poured himself some coffee, adding a dollop of cream.

"So, tell us about Apryl," Jake said as he began setting up serving bowls and spoons in a line along the counter.

"She's a bit of a wild child, you'll see. Her family owns a chain of hardware stores in the greater Charleston area. Harrington is her last name,"

John said. "We met down in Florida where I live, a bit north of Daytona Beach."

Abbey pulled the salad out of the fridge, mixed in some crumbled feta cheese, and set it next to the crock pot. "How'd that happen, may I ask?"

"Surfing," he said, with a chuckle. "She's a good surfer. Loves it as much as I do. We've been seeing each other for maybe four months now, but this is our first trip together."

They all looked up when they heard the other two coming down the stairs and watched Louie and Miranda sashay in together.

"Louie, my man, get some of this wonderful coffee," said John, holding up his mug. "But first let me introduce you to Abbey and Jake. Louie Stonebridge is a good friend of mine from school. We lived off campus for the last three years, with a revolving assortment of other characters."

"And this is my lady, Miranda Sanchez," Louie said.

"Hi to you good people," she said. Her dark hair was braided and brought forward over one shoulder. She wore gold earrings; her dark eyes twinkled under arched eyebrows,.

She also had a beautiful full-color cluster of red roses inked into her left forearm, and when she leaned over, Jake could see a matching but much smaller single flower on the soft upper curve of her right breast.

Jake just thought, *Wow*. "Nice to meet you, Miranda," he said, recovering. He noticed Louie had a similar earring to Miranda's, but only on the right side. "Gosh, Miranda, you've got the pair of earrings, but looks like Louie lost one of his," he said, smiling.

Louie replied, "You know—she used that joke on me the first night I met her, down at a bar in Fort Lauderdale. And no, before you ask, it wasn't anybody's spring break."

"Well," Abbey said, "let's get you started. Grab a plate from the table and come serve yourselves whichever breakfast dishes you choose. In the crock pot, Jake cooked up some Boston baked beans with molasses, very sweet. In the next bowl is a balsamic cucumber salad. Third in line is turkey bacon."

Jake pulled the platter out of the warming oven and folded a towel over the top.

"The juice on the table is pomegranate, one of my personal favorites," continued the lady of the house. "We have some fresh sourdough bread if you'd like toast with butter and jam, and can quickly heat up some oatmeal with brown sugar if that strikes your fancy."

By this time, they'd all grabbed their plates in a semi-organized fashion and lined up at the counter. Since Miranda took a moment to fill up her coffee mug, she ended up last in line.

"I'd like some toast, if I may," said Louie.

"Me, too," said Miranda.

Abbey pulled out the bread, dropped it on a cutting board, carved off some generous slabs, and dropped the first slices into the toaster. She suggested Jake get out some of the jams and jellies, and he was happy to oblige.

Miranda obviously had noticed Abbey was in the family way, and invited them both out to join her and the other guests at the table, if they hadn't eaten.

Jake said he'd gotten up first, so had already had breakfast, but he'd be happy to bring out his coffee.

As they were settling themselves around the table, Miranda asked Abbey the obvious question about her growing belly.

"I'm about six and a half months, looking forward to delivering in the middle of March," Abbey replied. She poured herself some juice, buttered up some of the toast Jake brought out in a basket covered with a white napkin, and chose the apricot preserve.

Miranda's voice softened. "I've got a four-year-old daughter staying with my mother, who picks her up from preschool and babysits her regularly when I'm at work."

"What's her name?" asked Abbey.

"Sofía Aurora Elena Fernanda Rodriguez."

"Sofía . . . I love that name," Abbey mused.

"We call her Sofie," she said, pulling out her phone to show Abbey and Jake her pictures.

Abbey exclaimed, "She looks just like you."

Except for the tattoos, Jake thought.

John looked at Abbey. "Boy or girl?" he asked, receiving the explanation from Abbey that Jake didn't actually want to know yet. The three were a bit surprised, but laughed when they realized he was serious.

"These baked beans are stupendous, by the way," John said, lifting his spoon. "Did you really cook them all night long?"

"Indeed he did," said Abbey fondly.

"I've got a kid, too," said John more softly. "His name's Ernie, he's three years old, and he stays most of the time with his mom in Daytona Beach, but I get him on weekends and longer stretches over holidays." He pulled a couple of pictures out of his wallet to pass around.

Jake changed topics. "So, George Mason University's at least twenty minutes from here. How'd you guys pick Dragonfly Inn?"

Louie responded first. "The reviews on this place were great, but as we mentioned last night, the deciding factor was your pair of level two chargers. We're both driving EVs now, riding that wave of popularity around the country, even if it started on the coasts. So I figured we'd find some really *simpático* people, fun to talk with."

"What kind of work do you both do?" John asked.

"I just started teaching in the engineering department at GW," Jake said, "and Abbey completed her fellowship in infectious disease at the same school and joined an infectious disease practice in DC. I am the one mammothly concerned about the degrading climate, so the focus of my research and teaching is all about redesigning the electric grid around efficiency and renewable energy."

"Were the EV chargers mounted on the garage out there your idea?" Louie asked.

"Yeah," Jake said, "but the irony is that we have yet to get a plug-in hybrid. We'd love to hear about your experiences."

Abbey cautioned him by touching his foot under the table. "First though, tell us about your school reunion," she said, buttering up some more toast and sending Jake for more bacon.

John took over. "This is sort of our ten-year reunion, which was supposed to be last year, but it got screwed up by Covid and monkeypox and RSV issues. George Mason is playing catch-up, so there are twice as many reunions happening this year, and obviously we get to lead off in January."

"So your next reunion will actually be in four years?" asked Abbey.

They nodded in unison.

"I was a history major," John said. "But go figure—once I got back to Florida, I set up a custom construction company. When the weather's clear we can sometimes see rockets taking off from Cape Canaveral down south, very cool. I've got twelve guys working for me and we focus on superinsulated and energy-efficient homes and multiplexes. Florida is working belatedly with better building codes now that we have a new governor, plus the legislature almost flipped in the last election.

"But even before those changes we thought long and hard about climate change. We prefer to build away from the coast, on the highest ground we can find in our county, and we make our work hurricane-tough."

Jake was nodding, his toes tapping on the floor.

"Course, if we're building custom, the warm coastlines are gonna be where people want their new houses, right?"

"Music to my ears," Jake said. "Sounds superb. But what's the elevation above sea level? You must be aware of subsidence pretty much affecting the whole state of Florida, right?"

"Definitely thinking proactively. You'll be glad to hear Volusia County has an average elevation of eighty-two feet and a maximum elevation over two hundred. So we stick to higher ground where possible. But if our customers and their architects want to build closer to the coast, sometimes we have to build on stilts, or design for a first floor built with concrete, or build masonry walls intended for storage and parking underneath the rest of the house."

Abbey shifted her gaze. "So, what did you do, Louie, in school and since then?"

Louie leaned back, finished chewing, wiped his mouth with a napkin, and said, "I got a degree in business, and a minor in Spanish, which has turned out to be incredibly useful in Georgia," he said, looking over and winking at Miranda. He was taller than his friend but not as tanned, with short dark hair and glasses like Rick Perry from Texas.

"Ah, *mi cariño*," she said, smirking.

"*Pues, mi amor*," he responded in kind, and she rolled her eyes.

Jake went to the kitchen to refill the pitcher with pomegranate juice, saying, "And after school, what happened then?"

"I work for a company called SK Battery America," Louie said. He wore a thick polo shirt and leather shoes. "One of the biggest battery recyclers in the country—"

Jake pulled his head up above the open refrigerator door with both eyebrows raised. "I know that company. That's really cool. Tell us about it."

"As you know, there's worldwide competition to build the batteries and motors of the future. A few US companies took some stabs at this a couple of years ago. But the results were not to scale. The materials extracted from big batteries, however incomplete, were shipped to Asia where most batteries get manufactured . . . or remanufactured."

"Aside from Tesla's gigafactories," Jake interjected, and went back to pour himself some more coffee.

"Yeah, sure, Tesla of course," Louie said, pitching his voice to follow him into the kitchen. "But the factory where Miranda and I work is just as big as anything Elon Musk has, as long as thirteen football fields, with three shifts working around the clock. The factory has all kinds of work positions, like engineers, operators, and quality control techs. Supervisors who oversee the robots working behind windows in clean rooms. That's where rolls of copper and aluminum are coated with nickel, manganese, cobalt, and graphite and made into book-sized battery cells."

"What kind of vehicles are these destined for?" Abbey said, holding one hand on her belly.

"The Ford F-150 Lightning, of course," Miranda replied. "I've driven one—piece of cake and quiet, too. The torque's equivalent to something like 580 horsepower—"

"You're talking about my truck, Miranda," John interrupted, "so let me explain."

"*Mansplain*, you mean," she growled. "Louie just taught me that word."

"Okay, okay," he said, "but you wouldn't believe the acceleration. From a standing start, it gets to sixty miles an hour in about four seconds. The battery capacity is so large it can function as power for tools at a worksite, partially recharge another stalled electric car, or back up a house in case the grid goes down. And we actually use it when we're working at remote sites. Some days, I even deploy some solar panels to really keep things humming."

"I've heard it said the Rust Belt centered around Detroit and the Deep South is being recast as the Battery Belt," Jake said, grinning and purloining another piece of bacon from the platter.

But John was just getting revved up. "Back in 2023, sales really started to take off as supply chains firmed up, the feds suppressed inflation to about three percent, and competition drove the lower-end EVs down to a range of twenty-five thousand dollars after rebates. With the bare-bones schedule of required maintenance and people getting fed up filling a tank with eighty bucks worth of gas or diesel, it wasn't long before about fifteen percent of new car sales became electric.

"It marked a tipping point. Fossil fuel car companies and all the oil and gas companies panicked, because the curve takes off exponentially around there. Game over for those bad boys. And I pity the poor schmuck trying to sell a used car requiring liquid fuels."

Jake held up a hand. "Hey, remember a lot of those guys have moved on to jobs in drilling for geothermal, or doing construction of offshore wind turbines. Lots of good-paying jobs there."

"The top-end Lightning is still about eighty-five thousand dollars," Miranda added, "but look at the competition from Volvo, GM, Audi, Volkswagen, and other companies. All these prices are comin' down."

"What do these battery jobs pay?" Abbey said.

Jake could see she was enjoying the repartee.

Miranda spoke up while still passing around some pictures. "Since I work in HR—you know, human relations—I calculate all these numbers. We live in Jackson County, a kind of rural stretch of Georgia between Atlanta and Greenville, South Carolina, where the average income is not quite thirty-three thousand dollars. So, base pay of over eighteen bucks an hour is significant, more than twice the minimum wage in our state. With two members of a family working, they can actually hope to buy a home and maybe put their kids through school.

"Both machine and production operators pull down eighteen dollars an hour; payroll assistants, twenty. Chemical engineers earn over seventy-four thousand annually. Accountants get eighty-six thousand. These are good jobs."

Louie sat up straighter. "Not to mention health, vision, and dental insurance, and meals provided in the company cafeteria. That food's plenty good, I'll tell ya."

"The mayor of our town, Commerce, lured SK Battery with a three-hundred-million-dollar incentive package of tax breaks, grants, and cheap property from the state and Jackson County," Miranda added. "SK is a subsidiary of a South Korean energy conglomerate, hence, the initials. The chief executive is Jun Yong Jeong, who often tells people to call him Timothy.[21] He says that the community has been warm and welcoming, and that he loves Southern cuisine.

"They finished their build-out in 2023 with about three thousand employees, and can now build 21.5 billion kilowatt-hours of annual battery capacity, enough to power over 430,000 new cars a year. Most of that's going into vehicles like John's Lightning, but also for Volkswagen's ID.4 EV." She had put her phone away, her eyes still shining.

Louie weighed in on the big picture. "In 2022, Ford and SK agreed to expand their joint venture and build three more battery facilities in Tennessee and Kentucky. And listen to this: The unemployment rate in our county is down to two percent. Two percent." He jabbed a finger straight up. "And ninety percent of new hires are Black."

"Shows the environmental movement, working hand in hand, can and should be about social justice," Jake said quietly. "Well done, guys."

He set his mug down and sat up straighter. "But you're leaving out a major dilemma. What happens when EV batteries reach the end of their service life? I know they can still function for stationary storage, like backup batteries for a residence or small business. But if we're going to be producing tens of millions of EVs, where are all the critical battery elements gonna be found?" *I think I know part of this answer, but let's see what these folks have to say.*

John spoke up urgently. "This is nothing but a good news story. The truth is that ninety-five percent of the elements inside a battery can be reclaimed and recycled at a profit. Ninety-five percent,[22] which is unbelievable if you think about it.

"I'll tell you a story to prove this. There's a company called Redwood Materials[23] based in Carson City, Nevada, that works with Toyota and other auto companies. Redwood has a cooperative agreement to help create a circular economy for the collection, testing, and recycling of batteries into raw materials to create a sustainable supply chain. Their ambitious goal is to expand into areas like battery health screening, data management, and remanufacturing—essentially the supply of battery materials throughout North America.

"It's not just that they can separate out the aluminum, copper, cobalt, graphite, lithium, and nickel. They can also remanufacture these materials back into anodes and cathodes. These two battery terminals—like the ones you can see poking up on top of an old-fashioned, twelve-volt, sulfuric-acid car battery—make up about sixty-five percent of the cost of lithium-ion batteries, and they can be refurbished and supplied directly to auto manufacturers like Toyota. Nothing is sourced from or routed back overseas."

"I understand this is way more than theoretical at this point," Jake mused out loud.

"Clearly," John said. "Even back in 2022, they were accepting more than six gigawatt-hours of end-of-life batteries for recycling, factoring out to about ninety thousand cars' worth, or forty-four tons per year, of batteries. By 2024, they'll have succeeded in hitting their target of one hundred gigawatt-hours, enough to manufacture more than a million new EVs a year."

Miranda edged in. "You've gotta understand this is not just about Toyota; we have partnerships with Ford, Panasonic, Protera, and Volvo. I opted in for the employee stock ownership plan, you know, an ESOP, and it's doing great."

Jake leaned over and put his elbows on the table. "I have a special love affair with the eighteenth century, so let me tell you about the history of mining in the United States. Back when there weren't as many states to be united, in fact, in 1872, only seven years after the end of the Civil War, when Ulysses S. Grant was president. It was when John Wesley Powell, the one-armed veteran and second and historically most influential director of the United States Geological Survey, was completing his second expedition down the Colorado River.

"Congress dispatched another party to map areas of the Southwest, with the intent of suppressing the native peoples and consolidating the American expansion accomplished in the Mexican-American War ending in . . . 1848. Most western territories were years away from statehood, even decades away in the case of New Mexico."

"Jake, come on, get to the point," Abbey said, giving him a severe look. She turned back to her table comrades. "You can see why sometimes I think this guy should be teaching history, not engineering."

Jake sighed. "So, in the era of Manifest Destiny, Congress passed the General Mining Law of 1872.[24] It helped build a nation puffing out its chest, but also contributed to the desecration of land and water. The Law to this day largely governs the mining on public lands for gold, silver, lithium, nickel, and other 'hardrock' minerals. This should sound familiar. Attempts to modify this legislation have been frustrated for over a century and a half.

"At this point, taxpayers do not get a penny in royalties for the minerals private companies extract from public land. It wasn't until the 1970s that mining companies were finally required to restore or reclaim these sites. Some 140,000 of these remnant, archaic quarries have been cataloged, with an estimated 22,500 posing risks to humans and the environment. But probably hundreds of thousands more unidentified mines—"

Louie interrupted. "It sounds like you're worried that if we don't clean up mining in the United States, the electrical revolution's gonna fizzle."

Jake shook his head solemnly. "Which would be a damn shame. Trout Unlimited used 2015 EPA data to estimate we've got over a hundred thousand miles of streams contaminated with heavy metals and/or acid, often located in watersheds supplying drinking water for communities and habitats for trout and salmon."

"How do you remember all this stuff?" Miranda said.

Jake snorted. "I've been prepping for one of my classes. Anyway, the law as it stands today allows any individual or company to explore for minerals on public lands. If they locate a resource, they can stake a claim for a measly $2.50 to $5.00 an acre, which gives them a 'right to mine'—all they have to do is actuarially project the profitable extraction of a deposit.

"In 2022, Congress proposed establishing a mechanism for tribal nations to petition federal land administrators—such as the Bureau of Land Management—for expanded protections from mining on public lands with established cultural and spiritual significance. Moreover, land supervisors would be empowered to deny permits if they were deemed to cause irreparable harm to environmental, cultural, or scientific values."

"I'm assuming you're talking about archeological or paleontological sites," Abbey said. "Or do you mean something else?"

He shook his head, then took another sip of coffee. "Over its first three years, the global coronavirus epidemic revealed a lazy and lousy set of policies, which had allowed us to become dangerously over-reliant on foreign supply chains. And in 2022, the Russian invasion of Ukraine underlined the requirement for secure sources of critical materials needed for the combination of clean power and electric vehicles—"

Peering at her phone, Miranda held up her hand. "Sorry. We're losing track of time." She glanced at Louie. "Didn't you say we had to be there by nine o'clock?"

Standing almost reluctantly, Louie said, "It's not like we need to catch a train, Miranda. These events never start on time, especially the first day." He touched her elbow. "*No ten miedo, mujer.* We just need to get out of here in about twenty minutes. Plenty of time to brush teeth."

John got up, too. "Listen, we're not done talking yet, so it's a good thing we have another couple of days here. Let's plan on focusing on some select models, like the two we have parked out front, at breakfast tomorrow. Plus, you get to meet Apryl, probably tomorrow morning, as we'll surely get in as late as last night. She had almost eight hours of driving to get from Charleston to George Mason, but she's only about two hours out now, having stayed with an aunt in Richmond.

"On the way back home she and I will caravan," John said, "and I'll stay a couple of days with her at her place. I took a week off from work, which is a little slow this time of year anyway."

"Lots of people to catch up with," added Louie.

They said their goodbyes, including Abbey, who had to get to work, with patients in two hospitals and more scheduled at her office in the afternoon.

Jake was used to the sudden quiet when everyone had finally cleared out of the house. But he was good company for himself. He turned on the radio for morning news while clearing the table, thinking he'd have to set four places tomorrow morning. Six if a late reservation came in for the second room upstairs, Quarterdeck. Always had to be prepared.

He got all the dishes washed up and in the rack by the sink, put some remaining food in containers for the fridge, aside from one last piece of bacon, and left a few big pots and pans to soak until later. It was a Friday, but he had a ton of stuff to work on at home, as usual. As always.

The snow seemed to have stopped, without much accumulation. But they could get a big dump this weekend. As a Minnesotan, that'd be just fine, even if he had to shovel some of it. He had plenty of experience running in the white stuff—gloried in it, even.

Ready for whatever was coming this way.

NINETEEN

The Zeitgeist Electric

For walk where we will, we tread upon some story.
—*Marcus Tullius Cicero*

The storm forecast shifted from Vermont and New Hampshire, dropping more than half a foot on northern Virginia. Jake got Abbey up early since they had four guests to prepare for. Abbey laid out settings on the dining room table facing each other, which made for easier conversation. Same plates and bowls, except with the addition of soup spoons for the chowder.

Jake had his apron on with breaded fish cakes in the frying pan, flipping them occasionally until each side had a delicate, crispy, golden crust. The aroma was intoxicating, and he sometimes thought that the smell alone should wake up guests.

So he said it out loud. "Amazing—that just smelling this doesn't wake people up, Abbey."

"Well, you know Virginia is for lovers, loverboy, so maybe that's what's keeping them upstairs. Besides, I have no idea when they came in. Did you wake up?"

"Nope. Slept like a log. Nothing like wintertime, sleeping under a thick quilt. Frost on the windows and all that. We're not into February yet, so no one is tired of the snow, or at least most aren't," he said quietly.

He looked out the kitchen window, seeing for the first time a third car that wasn't their own, the whole lot gracefully bedecked in snow. "Well, it looks like Apryl made it okay since they only took one car last night.

"Wait, the new car is one of the two that got to charge overnight. I didn't know she had an EV; they didn't say anything about that at breakfast, yesterday." He stared for a moment, fascinated. Then using the oven as a warmer, he put four more fish cakes in, sprayed a bit more canola in the frying pan, and carefully dropped in the next four portions of fish, slightly separated.

Abbey prepared more coffee, as Jake had made some serious inroads on the first pot, which she emptied into a hot-java carafe for the table.

"Looks like the chowder's coming along," Abbey said as she stirred it and then replaced the top on the crock pot. "Did you put butter on the table, honey?"

"I did, and the salt and pepper, all the usual stuff. I'm going to hold off on getting out the sourdough, to keep it fresher. Although, from the enthusiastic reception yesterday, I'm betting they're going to want some more. I'll get out a cutting board and the breadbasket just in case." Jake looked up at her as he opened the base cabinet. "What's on your schedule today, lady?"

"I have three consults at the med center, one of them another hemorrhagic fever patient. And I'm on call for consults all weekend, you know the drill."

Jake nodded. "I do indeed. I need to take the Orange Line into DC today to pick up a couple of reference texts. Most of our information is available online, but not everything. And after class I wanna find time to run again this afternoon."

"Wait, I think I hear stirring upstairs," Abbey said. They each halted for a moment, listening to the sizzle of fish cakes and murmuration of the vent above the cooktop.

They could hear footsteps coming down the stairs. A couple of the treads had a squeak, so jewel thieves were never a concern at Dragonfly Inn, Jake always joked.

She reminded him again to get out some oyster crackers for the chowder just before two people strode into the kitchen.

"Morning, John. How are you two doing? Hope you're hungry," Abbey said.

"Most certainly we are," said the tall blond woman with a broad streak of pink in her hair. "I'm Apryl, with a 'y,'" she said, and held out a hand to Abbey. The lady had a big smile, bangles on both wrists, and no visible ink work, but a pierced right eyebrow. She was wearing a button-up, orange-and-black flannel shirt, so no visible tan line.

"Apryl, I'm Abbey, also with a 'y.' We heard all about you yesterday."

"Oh, really," Apryl said, raising both eyebrows, staring down John until he had to chuckle.

Jake said hello as well and pointed out the coffee and mugs on the kitchen table. The two of them poured some and sat, John at the island, Apryl facing him from a chair next to the coffee. She added a teaspoon of sugar, then some cream.

Apryl eyed Abbey with a measured gaze. "So, I heard you're a doctor, you're pregnant . . . obviously, just look, but nobody knows what sex the baby is, right?"

"More or less," Abbey said with a serene smile, "though I may have some inside information, so to speak."

Miranda and Louie arrived next, and Miranda responded to what she'd just heard. "Nothing wrong with being smart, guys, as long as that isn't your only defining characteristic." She patted Louie on the rear.

"Remember what I said about Virginia," stage-whispered Abbey to Jake.

He changed the subject before they had a chance to follow up with a question. "Time to get this show on the road. Each of you needs to grab a plate off the main table and come in and serve yourself," he said, pointing to the counter on the right of the sink as he turned to open the oven. "First up, in the crockpot is a hearty chicken-and-corn chowder. And I'm bringing out the breaded fish cakes, and there's tartar sauce at the end."

Abbey threw in her two bits, pointing through the doorway with her chin. "Already on the table, you've got your dragon fruit nectar, which is supposed to be high in vitamin C and carotenoids, not to mention minerals and fiber."

"Wow . . . that's really purple," Apryl said, "with some small dark . . ."

"Seeds, Apryl," Miranda said, "I've had it before—it's wonderful, kind of a sweet, creamy taste with some nuttiness. Just try it."

"Okay, I will," she said, nodding vigorously.

As they all situated themselves and dug in, Louie asked what kind of fish they were eating.

"It's a whitefish, cod this time," Jake said.

Abbey reminded them there was more sourdough for toast. The two ladies wanted some, so Jake got that started. The two guys said they were certainly going to have seconds of chowder, and were told there was plenty more. The dragon fruit nectar was a hit all around, so Jake refilled the pitcher.

"Apryl," Jake said, "I think you should be up at bat first. You may not have heard from your friends here that Abbey and I are in the market for an electrical sort of car, so why don't you give us an earful about yours?"

"Having seen your car, Apryl, which is on the smaller side, I suspect it's a coupe. This kind of car typically has a lower sticker price and faster charging times. They may have a smaller battery pack, which would limit their range, but the trade-off is that they weigh less and use less power." He bent his head down to scratch his forehead. "I'm trying to remember what I just read the other day . . . that's right, there's a study that found doubling the vehicle mass increased the consumption of certified and real-world energy by sixty percent and forty percent, respectively."

Apryl set her fork down. "Jake, I've got to say, you nailed it. My range is 114 miles with a battery at full charge, so I need to stop for a juice-up in under two hours. But I'm almost always a city driver, and I do appreciate how easy it is to find a parking place. I'm driving a 2022 Mini Cooper Electric SE Hardtop and I *love* it. Kind of a cross between a coupe and a hatchback, with two doors and it seats four passengers. Though it helps a lot if the people in back are short.

"The MSRP was $29,900—"

"What's MSRP again?" Abbey said.

"Manufacturer's suggested retail price," John piped in. "But everybody knows this is just a starting point in negotiation, not where you wanna end up. What kind of tax credits did you get?"

"Seven-and-a-half thousand from the feds. But the only state subsidy available from South Carolina was a five-hundred-dollar rebate for installation of a level two charger, which I couldn't use, since the place I'm renting had already installed them."

"So," Louie said, "that got you down to something under twenty-six thousand dollars after wrapping in taxes and fees. Not bad for a car with less than half the maintenance costs and easy-to-find, cheap fuel." Apryl just grinned like the cat that polished off the cream.

"Any new safety features?" Jake asked, leaning back.

"Standard lane departure warning with steering wheel vibration, as well as pedestrian and front collision warning. But I hate the vibration feature—it's irritating—so I just turn that off."

"I checked," John said, "and the car is listed at 181 horsepower and 199 foot-pounds of torque, nothing like what I'm driving. But still, nothing to sneeze at."

Apryl sniffed. "Yeah, well, I don't run a construction crew or go off-road, so for me this car is a great fit."

Jake nodded. "Where do you usually charge?"

"I work in a hardware store with a fair amount of parking. We installed level two chargers like the ones you have—four of them. Main problem is getting people not to park in these spots unless they have an EV. By and large, though, I charge at home in my apartment building, which is pretty new. The lower-level parking has three of the same charging stations. But the electricians were smart and installed wiring for three more spots, so it'll be a breeze to install the other sites as the need arises."

"I'm going to back up Apryl here," Jake said. "A lot of emphasis has been on extending the range of new EVs, often targeting over three hundred miles. But I think it's misplaced, because most trips are far shorter. Aside from this road trip you're on, when was the last time you drove three hundred miles? Do you know what proportion of the time Americans drive thirty miles or less in a day? Any idea?" he said, looking around the table but seeing no takers.

"Ninety-five percent of the time we average about twenty-nine miles. Living in an era of perceived—but probably not real—scarcity of components for lithium-ion batteries, wouldn't the world be better off with two EVs with 150-mile range, instead of one car with three hundred? Or a couple of plug-in hybrids for families that almost always run on battery, anyway? Or working out a way to rely more on mass transit? Or a whole bunch of e-bikes?

"In 2022, when Gavin Newsom instituted a rule to ban by 2035 the sale of new gas cars in California, in fact, with interim EV targets of thirty-five percent by 2026 and sixty-eight percent by 2030, you could almost feel the parade start to roll down a hill.

"At least a dozen other states jumped on the bandwagon. But you could also feel the red states' pain. Because the attorney generals of seventeen Republican-led states sued to revoke the federal waiver that allowed California to craft this legislation, which would block this new policy.

"Unhappily, the case was heard before the US Court of Appeals for the District of Columbia Circuit,[25] considered the nation's second-most powerful bench after the Supreme Court. And you know the story from that point." His forehead furrowed, and his fingers tapped the table twice.

"I forgot to ask, Apryl," John said, seemingly trying to break the mood, "how many times you stopped to charge getting here."

"Just three times, scheduling it so I was ready for a meal," Apryl replied. "Level three chargers are really fast. Also, I certainly needed some juice by the time I got back here to the Inn."

"Louie," Jake said, "I've got a question for you. So you went to school at a pretty elite university in Virginia, and then ended up outside of Atlanta. Georgia has been deemed politically purplish for those folks who like to color-code these United States. What do you make of the talk about a New

South, and do you think wind and solar power, plus EVs, will change the political tenor of the state?"

"Hmm, funny you should ask. We do have two Democratic senators, I'll remind you. When I go out and drink beer with my buddies, we typically steer away from politics unless it's an election year. Sure, in a sense, being exposed to new breeds of energy and trucks will make the general public more accepting of technological change, and some locals—and I emphasize some—have come around to thinking and worrying about climate change.

"It's not at all as clear that their politics are changing. If that is happening, it would be from the cosmopolitan influence of the greater Atlanta area.

"One thing for sure, with my background in business, I just had to invest in what I know about. So, Rivian, because I know it's a solid company and I'm a proud owner of the truck, but also Harley-Davidson, because I've been riding a Harley for years. So if the stocks go up, I can earn back my vehicle purchase costs, since I'm nudging demand up."

Jake's eyebrows lifted. "Not an electric motorcycle? I understand they're making some nice dirt bikes."

"Jake, only someone who doesn't ride a motorcycle would even ask that question. It's about history, man, American history."

"Point made. Now tell us why you picked Rivian."

"They call these 'electric adventure vehicles,' with the tagline of 'preserving the natural world for generations to come.' But this is almost a credible statement, because all modes of transport should be electrified to accomplish getting carbon out of our lives, except for maybe long-distance flight."

"Not exactly a Model T for the masses, though, right?" Abbey said. "What's the starting price?"

"It was seventy-three thousand when I bought mine, though at least they've ramped up production so the wait for one is a lot less than three years, like it was in about 2020. Subsequently, the price has gone up significantly, but I don't bother to keep track of that."

"Have you gone off-road with it?"

"Couple of times on dirt tracks along part of the Chattahoochee River. Works just like any truck, like John's Lightning."

"Whadaya like about it?" Jake was almost ready to get started on dishes but didn't want to interrupt the discussion. And he knew Abbey could check on her hospitalized patients at her leisure today, and he could see she was still engrossed.

"It's got four electric motors, one for each wheel, and what they call a 'large' 128.9-kilowatt-hour battery, because I bought it in 2022. After that, Rivian started offering a smaller standard and a larger 'max' battery, coupled with a dual-motor option with six hundred horsepower."

Abbey shook her head. "Any smaller features that make it more comfortable?"

"Heated and ventilated front seats, whose importance I didn't realized until I started to experience them. Better than cabin-air heating and cooling. But there is an important point to be made about cooling, especially in an electric vehicle. We've all seen people in a parking lot in the middle of winter with the engine running to keep the heater going, or in the middle of summer to keep the air conditioner on—"

"Jeez," Jake said, "they might have something like a 150 horsepower engine idling just to keep comfortable. The world's least efficient air conditioner—"

"Ah, Jake," Louie said, smiling. "Not true in an EV. You're running off the battery, which you can fill back up with more sunshine."

Abbey smiled, too. "Nice, your conscience is clear. You can sit there reading while you wait for someone in the grocery store, with no noise and no fumes assaulting people as they walk by."

"No disgusted glares and muttering from people like me," Jake said.

"Next item," Louie said. "The quad-motor setup allows you to do a 'tank turn' where the two left and two right wheels rotate in opposite directions. You can spin in a circle, like an ice skater, just not as fast."

Miranda raised her eyebrows. "Well, how fast is it in a straight line?"

"It depends on whether it's towing a trailer. You can punch it and get to sixty miles an hour in 3.3 seconds. Rivian says you can tow eleven thousand pounds—that's over five tons—but the battery depletes a lot faster."

"What's the range if you're not towing?" John asked.

"EPA rating is 314 miles, but at seventy-five miles an hour with twenty-inch all-terrain tires, I've read testing has shown it only delivers 220 miles. The newer max battery is the biggest pack available, advertised with at least a four-hundred-mile range."

"Any last points about comfort?" Abbey said, shifting a bit awkwardly in her chair.

"I like the separate large displays for both the gauge cluster and the infotainment system," Miranda said. "The storage up front under the hood is called a 'frunk,' which is cute." She smiled at Louie.

Louie grinned back at her. "There's also inside storage behind the cab but in front of the bed. Like any truck, twelve cubic feet of lockable space out in back. What I like best are three 110-volt AC outlets and an air compressor built right in."

"John," Jake said, "does the Lightning have these features too?"

"The Lightning also has a frunk—but I won't tell you what we call it—with not three but rather four electrical outlets. And two USB chargers. Only thing the Lightning can't match is a built-in air compressor. I'll say, it does sound nice. But we always carry an air compressor to a worksite, so with an extension cord, we can get it right into the building under construction." He stuck his tongue in his cheek. "Six of one, half-dozen of the other."

"Who's going to win the truck wars?" Jake said. "Tesla?"

"No way, man," John said. "The Cybertruck doesn't even look like a truck. No place to attach a rack on top."

"Enough is enough," Miranda said, setting her coffee cup down and reaching her arms up for a stretch. "We've done it again. We're gonna miss the start of festivities unless we hit the road, Jack."

"Yeah, she's right," John groaned. "Louie, whadaya say we take my truck today and yours tomorrow? I'll just ask you all to throw in some gas money." Nobody bothered to respond to that old joke, but they all got up and headed upstairs, still talking.

Jake looked at Abbey and reached across the table so she could reciprocate. "You going in to the hospital now?"

"Yep, it's time. We have these wonderful folks here for one more breakfast tomorrow. But honey?"

"Yeah?"

"Let's see if we can think of something else to talk about tomorrow besides cars."

"A wise man once said a guy can get in the last word in any discussion, as long as it's 'yes, dear.'"

"Thatsa my boy."

TWENTY

The Ice Palace

*Ice contains no future, just the past, sealed away. As if they're alive, everything
in the world is sealed up inside, clear and distinct. Ice can preserve all kinds
of things that way—cleanly, clearly. That's the essence of ice, the role it plays.*
—*Haruki Murakami,* Blind Willow, Sleeping Woman

*The glacier was God's great plough set at work ages ago to grind, furrow, and knead
over, as it were, the surface of the earth.*
—*Louis Agassiz,* Geological Sketches

The days were getting noticeably longer now, but winter was still entrenched. Dark, convoluted clouds had paraded by all afternoon until dusk, like great China Clippers braving the run around Cape Horn. Professor Ligachev and Jake had just gone inside to confer about a few final details, but those still locking up bikes or otherwise getting ready to enter the building collided with the first freezing, outsized drops of rain.

"Hey, it's hailing," someone called out. The last few folks turned at the door to appreciate the wonder, but a hailstorm rarely lasted long and this one was no exception. The hailstones were translucent pearls, destined to melt away unceremoniously.

People inside were getting settled in, consciously or unconsciously, often in the same row and even the same seat as prior visits, opening notebooks or computers or peering into their phones. Jake, in a pale yellow shirt and tan corduroy pants, waited patiently for the room to quiet, then began as before, introducing Mikhail Ligachev for the benefit of the new people who filled the auditorium almost to full capacity.

Jake made an effort to project his voice, which Abbey was always encouraging him to do. "We're going to start with an examination of the cryosphere tonight—ice breaking bad—one of the principal manifestations of climate disruption. By the way, if I haven't mentioned this to you before, feel free to leave your phones turned on, with the explicit understanding you'll need to put any call on speakerphone and share it with all of us." A few people reached into their pockets. By the time he'd finished speaking, Mikhail had written on the board:

Secrets Locked in Ice

Mikhail had on a green bow tie tonight that matched the color of his rumpled trousers. He began with a personal story. "When I was a young undergraduate like many of you, well back in the last century, I journeyed to Alaska at the start of my summer break and ran the Midnight Sun Marathon in Anchorage. Then I took the train to Fairbanks, camped on the slope of Mount Denali, and ended up at Glacier Bay. I had nothing specific planned other than camping in the area, but I made the fortuitous discovery they had single- and double-kayak rentals. I chose a double so I could put my bulky pack inside the bow compartment, which was protected by the rubber kayak skirt cinched up over it. I made the arrangements to have their tourist excursion boat drop me off with my kayak and gear well up in the west arm of this fjord system the next morning, and then return to pick me up three days hence.

"I reveled in the only solo, three-day kayak I've ever done. The last glaciation over twenty thousand years ago had engulfed and sculpted Glacier Bay, which was graced by many remaining tidewater glaciers. There are fewer now. I was young and invulnerable, so I got up close enough to the calving ice that the waves would rock my kayak up several feet. I made sure to turn my bow into them to await their arrival. I fell in love that summer with gorgeous, gigantic ice scaling down to the crazy quilt of small pieces floating on the seawater, pieces I would later learn were called 'brash.' It fizzed with the susurration of champagne just poured into a flute. I guess back then I was pretty brash myself."

Jake just smiled, knowing how Abbey liked this speaker as much as he did.

"This would be an appropriate juncture to tell you how these glaciers are assembled from snow falling in flurries. As the snowfall packs down, it's initially made up of ninety percent air within and between the individual snowflakes. With the time and weight of subsequent snowdrifts, they pack closer together, disrupting the delicate crystalline architecture and leaving room for only half the now modestly compressed air. After granules, the next progression is to firn—spelled with an 'i'—at which point they are clumped down into a tiny, deformed fraction of their original size, with twenty to thirty percent preserved gas from the ancient atmosphere, as you can see in this graphic." He waved his arm to the right.

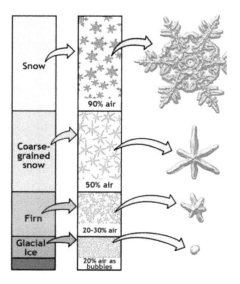

Figure 8: Image by Department of Geography and Environmental Science at Hunter College, used in "Glaciers," National Snow and Ice Data Center, accessed June 30, 2023, https://nsidc.org/learn/parts-cryosphere/glaciers.

"The final stage is glacial ice, with small pellets of residual snowflakes fused and frozen together around tiny pressurized bubbles relegated to only a fifth of the original volume. This slide has a nice summary."

Glacial Formation from Snow

- fresh powder   90% air
- granular ice   50% air
- firn   20–30% air
- final ice   20% enclosed [pressurized] bubbles

Data Derived from Glacial Bubbles

- pollen, ash, dust
- CO_2, or carbon dioxide   within few parts per million
- CH_4, or methane   within few parts per billion
- OCS, or carbonyl sulfide   within few parts per trillion

Global mean surface temperature determined by delta-oxygen-18 in bubble-adjacent ice.

"These bubbles contain not just primordial atmosphere, but also a wealth of information. Much of this treasure is older than the Holocene and Anthropocene epochs combined, even hundreds of thousands of years and more. This ancient air may contain tiny traces of pollen, dust, and ash.

"The trapped carbon dioxide can be measured to within a few parts per million going back eight hundred thousand years, at least in the Antarctic.

"Methane can be determined to within ten parts per billion.

"Carbonyl sulfide—a chemical that inversely measures global photosynthesis over an entire year—can be resolved to several parts per trillion.

"And if the ice is collected from an ice core with appropriate dating, then determining the level of delta-oxygen-18 in the bubble's surrounding ice—the ratio of oxygen-18 to oxygen-16—establishes how cold the world was during that year.

"What I didn't know then is that brash ice melting in Alaska was not the fizz of carbonation but rather the soft, sonorous sound of fossil air finally escaping its long confinement, with its wealth of information scattered to the winds forever."

As Mikhail posted, Jake realized anew that their speaker tonight had the soul of a poet.

Atmosphere + Clouds

"The atmosphere and its constituent clouds are a complicated part of this whole picture, for they can either protect ice or threaten it. As inhabitants of an inner planet, we get to experience water in all three phases: gas, liquid, and solid. From the bottom up, the atmosphere is divided into the troposphere, stratosphere, mesosphere, thermosphere, and exosphere, with an overlapping ionosphere itself divided into three sections. But we'll limit this discussion to the pertinent lower two layers as detailed on this slide."

Troposphere[26]

- height up to 14.5 km [9 miles] at equator, 8 km [5 miles] at poles in winter
- holds 75% of atmospheric mass
- holds 99% water + aerosols, hence nearly all weather
- tropopause -70°F or -57°C, dividing line with stratosphere
- tropopause freezes any escaping water to ice crystals

Stratosphere

- reaches from tropopause up to 50 km [31 miles]
- mirror image of troposphere with higher temps near top
- 97–99% UV absorbed by ozone, concentrated mid-level stratosphere
- constant winds exceed troposphere, up to 220 km/h polar vortices
- jet aircraft travel in lower stratosphere

"Gravity is the weakest of the four fundamental forces of nature, and would be unable to trap water were it not for the tropopause at the top of the troposphere exhibiting a barrier of minus seventy degrees Fahrenheit, which freezes any ascendent water vapor into ice crystals that return as 'fall streaks' of ice. Therefore, *almost* all weather manifests only in the troposphere trapped below this frigid boundary. Were this not true, our planet might have lost all of its surface water hundreds of millions of years ago." Mikhail let that information sink in for a moment.

Jake turned over a page and kept writing notes. *Damn, we could have ended up in stillsuits on Arrakis.*

"However, exceptionally energetic convection processes, such as billowing volcanic eruptions and 'overshooting tops' in intense supercell thunderstorms, may carry material up into the stratosphere on a focal and impermanent basis.

"Life is also highly dependent on mid-level stratospheric ozone blocking out almost all of the intense and destructive ultraviolet radiation, about ninety-eight percent of it, in fact. And the *steering* effects of fierce and constant winds above the tropopause drive storm tracks down in the troposphere below it.

"It must be emphasized that clouds—and their portrayals in climate models—are one of the most exasperating areas of this science. I begin with the description of the three broad groups of cirrus, cumulus, and stratus clouds, with the characteristics of cirrus clouds summarized on this slide."

Cirrus Clouds

- form in ascent toward frigid tropopause
- 10–100-fold less moisture than cumulus
- typically cover > 70% planet
- mixture of tiny water droplets + ice particles
- transparent or translucent to visible light
- opaque to infrared light or "heat radiation"
- warm planet both daytime + nighttime

"You'll note that cirrus clouds are often close to invisible. But how can you call it a cloud if you can't see it? Because instrumentation can still detect it as opaque to infrared radiation trying to escape the planet.

"Next up are cumulus clouds," Mikhail said, gesturing to his left.

Cumulus Clouds

- cumulonimbus if raining
- rise like hot-air balloons with focused convection
- moist air counterintuitively less dense
- daytime net effect cooling [shading]
- nighttime net effect warming [blocks IR escape]

"Wait, when it's hot and humid," someone interjected from the second row, "people say the air feels heavy. So how can humid air be less dense than dry air?"

"Good question," Mikhail said. "At a given temperature and pressure, only a certain number of separate gaseous particles can occupy any volume, a concept first articulated by a guy named Avogadro in the nineteenth century."

Jake had to chuckle. *Gotta love the 1800s.*

"If smaller, lighter water molecules displace double—or diatomic—nitrogen and oxygen, the summed weight of the volume will have decreased. It's counterintuitive, isn't it? But it explains why cumulus clouds convect vertically up, not down." He pointed first in one direction, then the other.

"Last cloud type is stratus. Notice that like cumulus clouds, these are thick enough to cool by shading in the daytime, but warm by insulating at night. If you think about it, these are common, everyday observations—all of them."

Stratus or Stratiform Clouds

- low-level clouds with horizontal layering + uniform base
- form when sheet warm, moist air lifts off + depressurizes
- light rain drizzle, if any
- daytime net effect cooling [shading]
- nighttime net effect warming [blocks IR escape]

"Climate modelers struggle with the impact of changing cloud cover, in particular, the uncertainty over the effect of warming on low clouds. A pertinent example of these clouds' importance is the Greenland ice sheet. Nocturnal cloud cover there is increasing, raising its temperature by up to three degrees Fahrenheit and contributing nearly thirty percent to accelerated GIS melting. In other words, the cloud layer functions as a thick quilt." Erasing the board, Mikhail wrote:

Cryosphere

"The warming cryosphere consists of multiple components, summarized in the next slide. Rising temperatures variably affect all of these. Only the melting of snow and ice sitting on top of land will contribute to sea level rise. Besides Greenland—the world's largest island—we focus on the Arctic, which is ocean surrounded by landmasses, and the Antarctic, a landmass surrounded by ocean."

Cryosphere

- ice sheet: large icecaps in Greenland + Antarctica, ice sheets restricted to 2 sites during modern interglacials
- sea ice: forms from snow on ocean, with melt + regrowth seasons, important variable in dynamically decreasing + increasing albedo
- permafrost + marine methane hydrates
- sink-to-source GHG, CO_2 + CH_4
- montane glaciers: Tibetan Plateau, Himalayan + Karakoram[27] ranges, sometimes called the "Third Pole"—after North + South poles
- seasonal snow-cover extent + duration critical in hydrologic cycle

---

"This next slide is the one that always gets to people, raising concerns about how much the ocean would hypothetically rise if all the ice melted. While we know that this would take centuries and centuries, we often update the information in the direction of faster rates as models become more sophisticated."

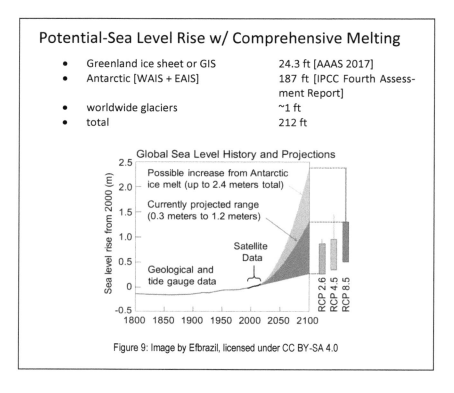

Figure 9: Image by Efbrazil, licensed under CC BY-SA 4.0

"Ironically, an ocean this much deeper would be the least of our worries if the world got that warm. But the actual territorial loss adds up to a couple percent of world surface area, land retreating from twenty-nine percent of the globe to perhaps twenty-seven percent, unfortunately leading to the loss of many of our most important cities." The crowd was pretty quiet as he quickly scribbled:

Aerosols

As he looked at these numbers being laid out so dispassionately, Jake could feel a rising sense of urgency.

"New topic," Mikhail said. "An aerosol by definition is an accumulation of fine colloidal particles, or tiny liquid droplets suspended in a gas or liquid. I picture this as something like tapioca pudding, my favorite. Clouds and fog of course are aerosols, but so are other substances. One of the most important is sulfate in fine particles, or droplets, of sulfate or sulfuric acid. The two main sulfuric acid precursors are sulfur dioxide from human or volcanic sources and dimethyl sulfide from marine plankton."[28]

Jake recalled that sulfate particles inhaled into the lung turned into sulfuric acid. *Bad actor.*

"All of these usually have a relatively short residence time in the atmosphere, barring the exception of volcanic ejecta vaulting up and piercing into the stratosphere.

"Another problem is coal-fired generators that release not just sulfur but lead, mercury, and carbon dioxide. But again, a completely natural source is dimethyl sulfide or DMS, the major substance released by dead or dying marine phytoplankton on a saltwater beach, creating the redolent smell of the littoral or seaside environment.

"A burning tree converts into about ninety percent carbon dioxide by weight, followed in diminishing order of importance by carbon monoxide, other gases, and ash. The minute particles of smoke from wildfires represent a more prevalent aerosol around the world, including from accidentally or intentionally set fires.

"Climate change is blamed for more than half of the increased fire danger in the western United States. Again, refer to the lengthy list of factors on this slide contributing to this risk, but fifty-five percent of the increased desiccation and a doubling of the area burned are attributable to climate disruption." Again Mikhail turned to review the slide.

Roots of Western US Wildfire Exacerbation

- changes in the Pacific
- historic fire suppression
- building into wildland urban interface [WUI]
- accidental ignitions
- accentuated lightning
- bark beetles
- declining mountain snowpack

Carbon Sinks Converted to Sources

- release heat + CO_2 [90%]
- reduce biota available to reabsorb CO_2

"Sea salt may be popular in upscale groceries, but it's also one of the most widely distributed natural aerosols. It's mainly composed of sodium chloride, but also salts of potassium, magnesium, and calcium, with a 'scattering albedo' as large as 0.97. It's produced by bursting bubbles as well as wind shear avulsion of wave tops. The sea salt wafted into the atmosphere reflects sunlight really well, enough that it saves us 0.6–1.0 watts per square meter of heating.

"Aerosols, such as sea salt, as well as dust, pollen, bacteria, and spores from wind erosion, will eventually settle out of the troposphere to the ground. They may act as cloud condensation nuclei and thus undergo 'wet deposition'[29]—essentially, raining out. Or they may respond to gradual gravitational sinking called 'dry deposition,' virtually the sole mechanism for removal from the stratosphere."

Jake hoped nobody was concocting any jokes about dry deposition. *Mikhail may be technical, but he's really a cool guy.*

"The 1991 eruption of Mount Pinatubo in the Philippines is a fine example of aerosol impact. This cataclysmic event ejected over a cubic mile of material into the atmosphere.[30] Downslope, an associated pyroclastic cascade of pumice and ash occurred. Some twenty million metric tons, or megatonnes, of sulfur dioxide were lofted into the stratosphere. The global influence was a 0.5 degrees Celsius drop in temperature from 1991 to 1993. This effect would have been about 0.7 degrees Celsius if a simultaneous El Niño event hadn't taken place. Mount Pinatubo altered the effect by a negative four watts per square meter, temporarily reversing global warming."

Jake scratched his head at this one as Mikhail wrote:

Sand + Gravel

"Let's talk about something more down-to-earth, namely sand and gravel. These substances represent the largest portion of primary material inputs to construction and industry in the world, almost eighty percent by weight. They are clearly the most widely extracted substances, far exceeding fossil fuels and biomass.

"In India, sand extraction has even led to civil strife, with a 'sand mafia' of powerful, organized criminal groups and hundreds killed in the 'sand wars.'[31] This level of competition creates heavy burdens for local populations, including farmers, fishers, and even those fetching water, typically women and children.

"The Mekong Delta of Vietnam is a critical food-producing region in Southeast Asia. Unfettered sand mining has led to enhanced subsurface salt-wedge intrusion during the dry season, plus damage to domestic water supplies and increased salinization of cultivated land. Even worse, extracting sand spreads infectious diseases, such as malaria from anopheline mosquitoes and leishmaniasis carried by sand fleas.

"This brings us to a set of infectious disease dilemmas. As snow and ice melt and temperatures and humidity rise in their stead, the global reach of ticks, mosquitoes, and other arthropod vectors expands. Some examples of disorders associated with these disease-carriers are summarized on this slide."

Vector-Borne Diseases

- anaplasmosis, babesiosis
- Rocky Mountain spotted fever, tularemia
- yellow fever [though existing effective vaccine]
- no protection for dengue, Zika, chikungunya virus group

Other Mosquito-Borne Diseases

- *Aedes aegypti* + *Aedes albopictus* mosquito vectors
- 4 species of malaria
- West Nile virus
- encephalitis viruses [eastern Equine, western Equine, St. Louis]

Globally over one million people die from mosquito-transmitted disease yearly

"This is a good time for a break, after which we are going to have Dr. Abbey London as our special guest speaker. She is going to expand our understanding of these disorders."

Mikhail left the slide up as people exited left and right to get outside air. Jake and Abbey stayed behind to talk with him, Jake feeling excited for Abbey.

TWENTY-ONE

A Delicate Dance

The very essence of romance is uncertainty.
—*Oscar Wilde,* The Importance of Being Earnest

When love is not madness it is not love.
—*Pedro Calderón de la Barca*

Rachel and Brian had wandered further outside than the others, having wisely retrieved her umbrella after pulling on their coats. They came together under a tree, enjoying the sleet hitting the leaves with splats and then dancing around their feet with tiny, staccato taps. He had his usual peacoat on, though he'd never been in the Coast Guard, and hunched his shoulders up to try and stop errant freezing drops from going down his neck. He had boots on but the legs of his cargo pants were getting wet.

Looking up as if for inspiration, he said, "Rachel," and he knew she could tell from the slow, serious cadence in his voice that something was up. He tried to hold the umbrella between them, switching it nervously from one hand to the other.

Slowly raising her eyebrows and looking straight into his eyes, she had a question in her face. He stamped his feet, trying not to make splashes. "Rachel, uh, we've gotta . . . we've got a really good relationship. I knew from the moment I first saw you that you were very intellectual."

"Old joke, Brian, you know that. Instead you recognized an angel standing before you in all her glory, right?" He recalled his keen attraction to her auburn hair and blue eyes that first day, but he had never told her he was fatally attracted to the tenor of her voice before he'd even turned around in the restaurant. Before he'd become enchanted by that angelic vision.

"Yes, damn it, but let me come up with my own lines." She tightened her lips, her face turning pink. She suppressed a smile, tilted her head to the side, and waited. "Well, you can't deny your intelligence. You're doing a double major, and you're really smart, we all know that." She waited a fraction more.

"I . . . I think we should move in together, Rach, I'm ready for that. We haven't talked about it, but maybe you've been giving me hints, saying how much you enjoyed staying over at my place in the carriage house. I could talk to my landlord; I'm sure she'd be okay with that. We could travel in together on the Orange Line, or bike in nice weather—"

"Not like today, in other words," she slipped in, and he stopped, uncertain.

Then he continued. "And we're both in the last year of our degrees, kind of on the same track as Jake and Abbey."

She shook her head but couldn't resist a smile at last. "I knew you'd bring that up. But you haven't really said why you want to do this, Bri . . . why now? You must have a reason. I know I do."

Brian thought he saw a way out of this impasse. "Well, that's good, tell me why you think we should do this, Rachel—"

Calmly and quietly she said, "I asked you . . . first."

Brian could see that she was resolute from her wide-open, unblinking eyes. He stuck to his courage, his face opening up, almost in surprise. "Because I'm falling in love with you." He saw her eyes shining. "I love you, Rachel. I need you, I want you. I cherish you."

She slid her arms inside his coat and hugged him. She was almost his height; they fit together well. He kissed her on the lips hard, then tenderly on the cheek, and nuzzled her neck.

"And I love you, too, you big oaf. Took you long enough to figure this out. And yes, I do think we should try out some cohabitation."

TWENTY-TWO

Eemian Warning

Mikhail waited as a few stragglers made their way back to their seats, then got started without Jake's help this time.

"Dr. London's not British, though she was born in India because her American parents were stationed there. And she recently completed a fellowship in infectious disease here at the medical center. She's going to tell us about infections expanding around the globe."

Jake was proud of how Abbey marched down the steps and came to the center, standing closer to the front row than the other speakers.

"While I don't have any prepared comments, I most certainly have formally presented data on some of these so-called invasive or emerging diseases, and I've had the good fortune to consult on a fascinating array of patients.

"Rocky Mountain spotted fever and anaplasma are caused by small bacteria of an unusual group called rickettsia,[32] miniscule bacteria typically transmitted by ticks. I've been involved in the treatment of several cases of RMSF and one of anaplasma. These diseases are rare but become serious if diagnosis is delayed. RMSF is characterized by fever, headache, and rash. It's treatable with antibiotics, but again, it can be fatal so it requires an urgent diagnosis.

"Tularemia is caused by a bacterium called *Francisella tularensis*, usually acquired in association with rabbits or rodents. But I have yet to have personal clinical experience. This is another uncommon disease and its most serious manifestation is pneumonia." She turned to look at the slide again.

"Yellow fever does have an effective vaccine, which is recommended only for travelers to certain countries in Africa and South America. This viral disease itself causes jaundice—hence the name—and it can certainly be fatal. The vaccine is quite effective, although carries risks in individuals over sixty-five, as the vaccine is an attenuated live virus.

"I've treated a number of patients with dengue fever, commonly known as 'breakbone fever' because of its associated intense pain, and also another agent in the same viral group, namely chikungunya. Historically, dengue had retreated from the United States to Central and South America, but demonstrated that it can return with the appropriate carrier—namely the *Aedes aegypti* or *A. albopictus* mosquitoes. In fact, it has reappeared in Hawaii and parts of our Southeast.

"I've not yet seen Zika, but I am aware of several cases in New Jersey in travelers to endemic areas. Zika is particularly concerning because of the initial cases of microcephaly in pregnancies complicated by infection," she said. Jake saw her unconsciously holding a hand over her abdomen. "Even transmitted by the fathers of these babies well after the men had left an endemic area, so this is another addition to the lengthening list of sexually transmitted infections.

"Malaria is a fascinating parasitic infection and important worldwide. Five *Plasmodium* species attack humans, but only three can be chronic and recurrent, and I've treated more than a handful of those cases here in DC. Malaria was an endemic disease in the capital back when Europeans settled this country over the objections of the eastern tribes.

"Data published in 2023 show the anopheline mosquitoes, which carry malaria, have moved to higher elevations over the last century by roughly twenty-one feet per year, and away from the equator by about three miles per year.[33] But the rate of migration may be increasing, along with the acceleration of warming over the last few decades.

"To answer your basic question about whether changing vector ranges will put our population at risk for novel diseases, the answer is indisputably yes, and that is further reason to stop and reverse altering the climate. These diseases are no longer just tropical and exotic."

"Was the coronavirus just a surprise from out in left field?" Mikhail said.

"On the contrary, COVID-19 was actually predictable because of a multitude of viral, bacterial, and parasitic infections—often enzootic—waiting to break out from animals into the human population. We're getting better at viral surveillance in places like South Asia and Africa. But once this enzootic coronavirus infection crossed over into humans, its respiratory spread took on a life of its own, without any insect or arachnid—by which I mean, tick—vector required. That's why I didn't bring it up."

Mikhail turned from writing on the board. "Thanks, Abbey, though I suspect some people may have further questions afterward.

"Next, let us discuss the storm clouds on a trajectory toward us."

Intense Storminess

"An article in *Nature Reviews Earth & Environment* in 2020 heightened interest in studying mesoscale convective system storms,[34] which even then were causing economic losses in this country exceeding twenty billion dollars a year. These are huge thunderstorms having a diameter of a hundred kilometers or more, which have increased in frequency and intensity over the last thirty-five years, along with tornadoes, hailstorms, and derechos. Fatalities and economic losses have resulted, and the projection is that these events will triple in North America this century.

"This is only one piece of quantitative evidence for accentuated storminess per degree of warming. Hurricanes are another part of this picture. Though we still lack evidence for an increase in the total number of hurricanes, even after the impressive 2020 and 2022 seasons, we have enough experience to claim that more severe storms are already arriving. These are classified according to the Saffir–Simpson Hurricane Wind Scale.

"Western Hemispheric tropical cyclones—those exceeding the intensities of tropical depressions and tropical storms—are slotted into one of five categories distinguished by their sustained wind intensities. But this is only a single and limited metric based on wind speed, while other measures—more insightful, perhaps—look at the intensity of precipitation and maximum area of the storm."

Jake reminded himself that he'd always wanted to experience a hurricane. *Crazy, but there it is.*

Returning to the board, Mikhail wrote:

Greenland Ice Sheet

"Time to return to the world's largest island and the Greenland ice sheet, or GIS. It comprises unimaginable amounts of water on either side of the freezing point. NASA's project, Oceans Melting Greenland—one of the best acronyms I've ever come across . . ." It was only a beat or two before the laughter began.

"OMG[35] was mapped with state-of-the-art soundings by ships and other data sources, using penetrating microwave—radar, essentially—to map the substrate and ocean floor adjacent to and beneath Greenland's glaciers. A large, central part of the island bedrock lies below sea level, and this observation extends to several outlets reaching the open sea. The implication is that Greenland is melting both on top and from warm seawater underneath the bottom. At least fifty-five percent of the ice sheet's total drainage area is at risk from warmer, subjacent Atlantic water.

"Recall the Greenland ice sheet represents more than twenty-four potential feet of sea level rise, albeit on a geologic time scale. Centuries, in other words."

Rachel raised her hand. "So, if I get this right, you're saying over half the land surface of Greenland, the bedrock, is underwater?"

"Yes," Mikhail affirmed. "This is exactly the case. And the top of the ice is also melting. Melt ponds proliferate as we enter summer, which wrecks the local albedo. The blue of a pond enhances the absorption and deeper penetration of ultraviolet and visible light, though a shorter depth of infrared light, with resultant heating—"

Rachel decided that this comment begged a further question. "Why doesn't infrared penetrate as deeply as UV and visible light?"

"Recall water molecules are potent greenhouse gases. So infrared is easily absorbed not just by water vapor but by either liquid water or snow or ice, with significant warming as a result."

Rachel persisted. "So IR fails to penetrate far because the water absorbs it so avidly?" Mikhail assured her she was correct, but he paused, looking at her and the others to be sure they grasped the concept.

"Fresh snow, on the other hand, has an albedo of 0.8 to 0.9, with 1.0 being a near mirror-like surface. Bare white ice ranks as 0.7 and the open ocean and melt ponds plummet down to a value of merely 0.1."

Someone in back interjected. "Professor, you're saying blue ponds and oceans absorb more than ninety percent of the solar gain? Almost all of it?"

Mikhail nodded and continued. "These ponds may abruptly drain by a moulin, a sort of internal waterfall, which cascades down to bedrock and tends to lubricate the ice and quicken its pace of descent down the slope. More recently realized is that the potential gravitational energy of the water is converted to heat as it plunges through crevices, or moulins, in the ice." He was gesturing with his hands, like a TV weatherman.

Jake had to smile as he visualized this image.

"Most river deltas around the world are diminishing due to sea level rise. But in Greenland, fjords display expanding outwash deltas in comparison to photos taken by American pilots in the 1940s. Clearly, anything darkening the ice tends to increase the melting, which in turn builds these deltas. Finnish-Swedish explorer Nils A. E. Nordenskiöld suggested nearly 150 years ago that curious small, dark pits in the ice were partly responsible for melting, and again, we see that some of the basic research on climate was accomplished in the nineteenth century."[36]

Jake grinned in appreciation.

"The color of what is now called cryoconite in these pits comes from dust, soot, bacteria, and microalgae.[37] The soot may represent smoke from wildfires a continent away. And the microalgae produce a dark pigment to protect themselves from the harsh ultraviolet in the same way we might develop a tan. But logically, these pigments further diminish reflectance.

"Okay, time to head to the other pole," he said, erasing and replacing Greenland with Antarctica.

Antarctic Ice Sheet

"Almost all of these observations hold true for Antarctica as well, except for the deltas. But the scale is colossal. The bedrock of this continent looks more like an archipelago than a contiguous landmass. The Antarctic Peninsula extends up toward Tierra del Fuego at the southern tip of South America. If this ice starts melting in quantity, then the whole land surface will experience postglacial isotactic rebound—"

Another student on the right, a guy with a baseball cap, interrupted. "What is that . . . iso what?"

Mikhail replied, "Sorry, good question. Ice is very heavy, as anyone knows who's ever carried a bag of it home for a summer barbecue. When several kilometers of ice have compressed a broad area for hundreds of thousands or millions of years, that pressure is released by the melting. The subjacent land surface will then spend centuries rising back up, like slowly leavening bread dough.

"The ice sheet is divided into the West Antarctic ice sheet, or WAIS, and the East Antarctic ice sheet, or EAIS. The spine of the Transantarctic Mountains separates the two. Immediately adjacent to the mountains to the west is the great Bentley Subglacial Trench,[38] which plunges more than a mile and a half below sea level, the deepest place in the world not under an ocean. But the ice above the Trench extends more than a mile above sea level all the way down to that base.

"The bedrock below WAIS is generally downsloping from the ocean edge toward the inland part of the continent. Either tidewater glaciers or ice sheets often extend out in a contiguous, floating ice shelf. The border between fully grounded ice and the floating shelf is called the grounding line, and if the bedrock is downsloping, then the boundary gets deeper and deeper the further into the continent it migrates with melting.

"In 2020, a joint British and American team drilled down six hundred meters, or almost two thousand feet, to a portion of the eighty-mile-wide

Thwaites Glacier not far seaward of its grounding line.[39] What the instrumentation found down the borehole was profoundly disturbing. Tiny bits of ice were desquamating off the glacier and swirling around in the currents there. But most worrisome was the saltwater temperature: two degrees Celsius above freezing. The subjacent, warmer circumpolar water is actually melting the ice sheet from below." Mikhail wrinkled his forehead and was quiet for a moment, then resumed walking.

"Even more concerning is that subsequent measurements by robotic submersible revealed water temperatures in the same area of three to four degrees Celsius above freezing, which the researchers found extraordinary and ominous." He stopped pacing.

"The Thwaites Glacier is one of a set of six in the Amundsen Embayment in West Antarctica, the second broadest of these being Pine Island Glacier. If the buttressing effect of these glaciers is lost, like a cork in a bottle, then the ice in the deep subsea bowl will present vast, sheer cliffs above and below the waterline with worsening structural instability.

"To a geologist, water ice is no different than any other rock, aside from having a lower melting point than limestone or granite. If retreat is sufficiently relentless, then ice cliffs thousands of feet high above and below sea level would far exceed the water ice fracture limit, and we could potentially see destabilization over several decades and sea level rise of as much as eleven feet. Similarly deep Wilkes and Aurora basins beneath the East Antarctic ice sheet could experience hollowing out from below and lead to additional sea level rise at an alarming, combined rate." Mikhail walked a few feet back, then wrote:

Eemian Epoch

"The obvious question that presents itself is whether or not something like this has ever happened before. The most common comparison is to the Last Interglacial, the Eemian epoch, which stretched from 130,000 to 115,000 years ago. That was the last interval when Earth was as warm as it is today.

"The Eemian followed the Saale Glacial Stage and preceded the Weichsel Glacial Stage in Europe. Eemian was the name chosen after a stream of the same name, in Belgium, I believe. In either event, the Eemian was concurrent with Marine Isotope Stage 5e.

"Hippopotamuses roamed as far north as the Rhine in Germany and the Thames in Britain. Hazel and oak trees grew as far north as Finland. Robust geologic field data indicate the sea level rose three to four meters higher than now, possibly concluded by a rapid end-Eemian rise of six to nine meters higher than the present level."[40]

Brian whispered to Rachel, ". . . goddamn incredible . . . twenty to thirty feet higher . . .," while she just shook her head.

"Evidence in Bermuda and the Bahamas shows physical confirmation of storminess persisting over centuries, with hurricane strength exceeding anything in the modern era.

"Preserved regional sedimentary and geomorphic features attest to a turbulent end-Eemian world. Part of the evidence consists of overt wave-produced run-up on the windward side and chevron deposits on the lee side of low-lying islands such as the Bahamas.

"On the island of North Eleuthera, several enormous limestone boulders were plucked from seaward mid-Pleistocene outcrops and crashed onto a younger Pleistocene ridge twenty meters high. The weight of these megaliths is estimated at a thousand metric tons apiece. The associated storm surge would have to have been sixty-five feet high." He paused for emphasis.

"Think about that for a moment. We're accustomed to storm surges of eight or twelve feet. But not sixty-five feet. The huge boulders and the surface they landed on were absolutely different kinds of rock, indicating the boulders were transported by a huge storm. These findings were described in a seminal article by Hansen and his collaborators in the journal *Atmospheric Chemistry and Physics*. About eight or nine years ago, as I recall.[41]

"Contemplate the possible and relatively rapid rise at the end of the Eemian epoch to nine meters above our current ocean. That's almost thirty feet. We reached temperatures comparable to that stage in 2017, and have continued to exceed that mark. I'll reiterate that carbon dioxide is now higher in our atmosphere than it's been any time in the last six million years.

"Again, let's look at this logically, another *gedankenexperiment, or* thought experiment." He stopped, for one last time, to write:

Summary

"The first control knob with regard to climate change is heat-trapping gases, and that's where we have the greatest challenge and the most leverage.

"The second control knob is orbital cycles, and these cycles serendipitously may be on our side, for were it not for our releases of trace gases, the planet would be heading toward a cooler climate over the next several tens of thousands of years.

"The third control knob is the weathering of carbonate rocks, a process which unfortunately offers no practical way for influencing it, and what's more, the slow pace of weathering would require hundreds of thousands of years."

Not that we have that kind of time, Jake thought.

"The fourth control knob is tectonic plate motion, which presents two flaws from our perspective. First, humans cannot affect the magmatic circulation moving the continental and seafloor plates around. And second, this geologic unfolding requires not hundreds of thousands, but millions of years."

"While my ancestors were German and Russian, I do not believe we face *Götterdämmerung*—most often translated as *Twilight of the Gods*—the last of Richard Wagner's famous cycle of four musical dramas titled *Der Ring des Nibelungen*, or simply *The Ring*."[42]

"But even *Götterdämmerung* is a translation into German from the old Norse phrase, *Ragnarök*, which in Norse mythology invoked a prophesied final conflict between the gods and other entities.

"In this war, the planet was ultimately subjected to burning and immersion in water, but then, the renewal of the world." He stood facing them in silence for several seconds, concerned, Jake guessed, at the distraught expressions on some of their faces.

"Remember, we're a tough and imaginative species. And we've walked this planet for possibly as long as three hundred thousand years. We've done so with much more limited tools at our disposal: After all, humans survived several cycles of glaciation and deglaciation, including the most recent Eemian epoch with its hurricanes, of which we have no species memory. Certainly, no civilizational memory." Again, he stopped and appeared to be brooding.

"I've laid out a set of the problems we face. Next session, Max Baerbock and I will finish off the remaining challenges involved in repairing the climate system. We have no choice but to accomplish mitigation, adaptation, and withdrawal of carbon dioxide from the atmosphere and ocean."

Then, he smiled and declaimed firmly in his baritone, "So, let's get started." The applause was tentative at first, but then grew and was sustained, and people began standing to clap; ultimately everyone joined in.

Jake afterward went to gather up Abbey and their other friends, saying, "How's that for an operatic conclusion?"

"I expected a full-bodied opera star in a suit of Norse armor," Brian reflected, "holding a last long, plaintive note."

And Rachel finished. "I think we just heard that."

TWENTY-THREE

Homecoming

Home is where my habits have a habitat.
—*Fiona Apple*

Jake checked the thermometer mounted outside the kitchen window above the sink. Twenty-three—colder than when he'd gone to bed last night. The icicles below the gutters had lengthened, forked, and merged with quiet and patient abandon. The roof-mounted anemometer and weathervane showed the wind to be twelve to fifteen knots from the northeast, churning restlessly back and forth, so typical for this time of year. He remembered maximum snowfall peaked within a degree or two of freezing. *Where had he learned that again?*

He put on his heavy coat, braved the blast on the front porch, bare-handed the banister—because the steps had a thin veneer of fresh snow on top of an icy layer—slipped but didn't fall, twice, and made it down to ground level. From that point, he walked confidently, hands in his pockets. With a deep breath, he could feel the cold curl the hairs of his nose. The *New York Times* and the *Washington Post* were individually wrapped in butcher paper to keep them dry. The snow wasn't wet, so—no call for plastic.

With a slight frown, trudging back, he realized he'd have to carve out time to shovel their visitors out. Only one car was parked in the circular gravel drive in front of the house, with all of its windows covered and both windshield wipers angled out like the wings of yellow jackets, in this case to avoid them freezing in place. He tried to tread in his own footsteps upon return. He hadn't seen the car when Alice and Bob came in late last night, and now he wasn't even sure what color it was.

Shrugging off his coat in the boot room, he sat and pulled off his footgear to avoid tracking snow in the house and slipped on a pair of worn moccasins that had followed him since college in Minnesota. Knowing Abbey had her rounds with patients at the hospital, plus office hours, he sneaked into their bedroom on the first floor. She was sleeping on her left side, rib cage expanding and relaxing smoothly. He pulled the covers over on his side of the bed and slipped behind her back, spooning. He slipped his hand under the flannel nightgown to caress her belly, hoping to feel the baby kick. *How do women ever sleep with all the gymnastics going on inside?*

"What time is it?" Abbey murmured softly, without even opening her eyes.

He reached up to glory in her full, maternal breasts, but she said again, "Seriously, what time is it, Jacob? Stop that." She glanced at the clock, moved gracefully to a sitting position as she yawned, and asked him if he had breakfast ready for the guests.

Sighing, he said, "No," bowed to the inevitable, and left as silently as he'd arrived, the thief of her heart, closing the door with hardly a creak of the hinges.

He extended his interlaced fingers in front of him when he got to the kitchen, then got to work as soon as he turned on the oven to heat up. The table against the wall he set for two, with large plates, bowls for cereal, cloth napkins, and silverware. He looked at the orchid vase, but shook his head, leaving that management to the lady of the house.

Having prepared the dough for pumpkin-walnut muffins the previous night, he now applied canola oil in each cup of the pan and inserted spoonfuls of dough to just below the brim. The only juice selection was vintage apple, but that went down pretty well in wintertime—heated up with a cinnamon stick. Coffee, of course, a dark bean from Guatemala because of his Peace Corps history there, ground the night before so as to not wake sleeping guests in the morning. The yogurt was vanilla, but he put out a bowl of blueberries for mixing in.

When the oven was ready, he slid the muffin tray in and set the timer for thirty-five minutes. *Still no signs of life upstairs.*

He could make oatmeal in his sleep; it was a staple in his diet and none of it ever got wasted, even if guests didn't partake. Final step was getting out the square, ridged skillet, applying some more vegetable oil, and laying out the strips of turkey bacon. Both he and Abbey found this sumptuous, so even if guests turned out to be unexpectedly vegetarian, all was put to use. *Waste not, want not,* he thought, but he remembered that aphorism was predated by "wilful waste makes woeful want" from sometime in the sixteenth century. *Not much used, anymore.*

Abbey strode in grandly, wearing a blouse as blue as a robin's egg and a single simple pearl necklace with matching earrings. She served herself oatmeal and a separate bowl of yogurt, adding in some blueberries. Jake heated up a mug of apple juice for her in the microwave, though she opted against the cinnamon. In wintertime, the kitchen was the warmest room in the house in every conventional sense of the word.

Jake had told her only that their guests were Alice and Bob, anticipating a surprise for Abbey.

"Morning, Jake," Alice said when she arrived first. "Bob will be down shortly, but I need coffee."

"Abbey, this is Alice Schmidt, whom I met last night," Jake said, and the two women smiled and said hello. Alice appeared lithe and fit, possibly in her late thirties, with long, loose dark hair. Jake appreciated the musicality of her voice at least as much as he had last night.

Abbey added, "You can have a seat at either end of the table," as Jake poured a mug of coffee and set it next to her.

"Looks like you're in the family way, Abbey," Alice said. "When are you due?"

"Not until the middle of March, so I'm almost seven months now."

"Do you know if it's a boy or a girl?" But surprisingly, the answer came from Jake.

"She does know, but . . . I didn't want to know, so she's promised to not tell anybody ahead of time. All I know is the sonogram and all the tests have been normal. Not wanting to know meant I couldn't be there for the sono, but I was willing to pay that price. I'll just wait to see that video on my own time. Guess that makes me a traditionalist of sorts."

Abbey just gave her mysterious, Mona Lisa smile.

"Alice, I should let you know there're some pumpkin-walnut muffins in the oven that need about another ten minutes," Jake said as he turned over the bacon in the pan. "It's been our experience that even dyed-in-the-wool vegetarians succumb to temptation when traveling. And it's from a turkey, not a pig."

"Bob and I are definitely not vegetarians, Jake, so no worry about us on that score. We're here because Bob's never seen Washington, DC, before and I wanted to play tour guide."

Abbey had grown curious. "And do you and Bob have kids?"

"We do, a boy and a girl. David's eight and Clarissa's five. Their grandmother in Bryson City is taking care of them for a couple of days."

By this point Jake had put a platter of bacon in the center of the table and ladled out oatmeal into Alice's bowl. "Tell us about Bryson City," he said. "I hate to show off my typical American ignorance of geography, but what state is that, anyway?"

Her answer was North Carolina. "I grew up there. Classic, small-town Americana, only about 1,500 people, eighty-nine percent white, most of the rest, Native American. Nestled in a valley leading to some of North Carolina's most incredible outdoor places."

Jake could hear the smile in her voice.

"Other small towns are within striking distance, including Whittier, Cherokee, and Almond," she said. "And four rivers run through Swain County: the Nantahala, Oconaluftee, Tuckaseegee, and Little Tennessee."

"So, not Tuskegee as in Alabama, but 'Tuck-a-see-gee'?" Abbey said.

"Absolutely, ya got it in one. And, Bryson City is the county seat. Makes us pretty important in our neck of the woods."

"With a town that size, do ya sorta know everybody?" Abbey smiled as she asked.

"I could just say sho'nuff, but actually my parents were from Kansas City, then moved to Bryson City postretirement, so my accent is more midwestern than Carolinian. My mother had been in the Coast Guard, stationed at Charleston for a couple of years, so she persuaded my dad to retire in the Carolinas. But he died only a year later—never had a chance to meet his grandkids."

"I'm sorry for you, Alice, and for the kids not getting to know their grandpa," Abbey said softly. Even with all her infectious disease expertise, sometimes condolences were all she could offer a family.

"Thanks, Abbey, but that was over a decade ago. I tell the kids lots of stories about him and show them all the pictures."

"So, what do you do there?" Jake asked.

"I'm the proud proprietor of the Coffee House Caboose, situated right across the square from the train depot. We get a lot of tourist traffic from the train."

"And what does your partner do?" Abbey said.

Just then, Alice heard someone coming downstairs. "I think I'll let Bob explain that to you."

Into the kitchen walked Bob, and Abbey's eyebrows rose slowly, just as Jake had suspected. She looked to be the same age as Alice and just as fit, but half a head taller, with a deeper voice and a short, blond bob cut. The accent was Scandinavian when she finally spoke. "My name is Robban Ahlgreen. I'm from Sweden. Before you ask, I have played volleyball, but never on a beach. Never as a professional. My nickname is Bobby. Or, Bob."

Jake was enjoying this to no end. "Of course, you look a lot like a volleyball player, so you must hear that all the time." He was trying to look innocent in case Abbey glanced at him.

"Can you tell? Such a cliché, don't you think. But I am used to it."

"And, what kind of work do you do, Bob?" Abbey said, without missing a beat.

"I'm the town manager, reporting to the mayor and aldermen. Of course, my staff and I do all the real work of the local government." A satisfied smile rose on her face.

Abbey followed up. "And how did you two meet?"

Bob answered first. "I was at the University of North Carolina at Chapel Hill studying city and regional planning, with a second concentration in American studies." She looked over at Alice and said, "And this lady ran a coffee shop just across the street in Chapel Hill. So I became an even more serious—what do you say?—imbiber of coffee.

"Once we realized we were serious about each other, it became clear we were both looking for the same thing: a small-town experience somewhere on the East Coast," Alice said, picking up on the story as she crunched loudly on bacon and followed up with a gulp of her apple juice. "I was twenty-nine at that point, only a couple of years younger than Bob. My mom was incredibly supportive. She's even housing us as we're building our home, and she cosigned on the construction loan. I'd been saving up for years, and Bob had some significant family support."

Jake suspected there might be some inheritances down the line.

Bob smiled at Abbey meaningfully, so Abbey obliged her. "We decided to try for a baby almost right away, and Jake was fully on board, so to speak."

The three women laughed and he tried to hang his head, feigning shyness. But they weren't having it, so he laughed too. *Aw, shucks.*

Just then, the timer went off. Jake opened the oven, and an irresistible smell of baked pumpkin wafted across the room. He pulled out the tray and set it on the cooktop to cool down. His mouth watered and he suspected the smell did the same for the others. He set out some butter on the table and picked up the oatmeal and yogurt dishes and bussed them to the sink.

He pulled on his apron from the hook by the refrigerator. He knew the two women would share significant glances with Abbey about how he seemed "well-trained." *At least I'm used to it.*

"So, enough about babies," Jake said, turning back to the sink. "I wanna hear about this house you guys are building. I should mention I'm in the engineering department at GW, with a special focus on studying the electric grid."

"My parents are both architects and helped us a lot with design," Bob said. "Have you heard about the Passive House or *das Passivhaus* movement?"

"Indeed, I have. I wish my friend Brian were here. My degree is in electrical engineering, but his is structural engineering, and he's a real geek on this stuff."

"So, the Passivhaus movement began in Germany, a cloudy and temperate nation, so people vaguely familiar with the idea assume it's not for tropical or subtropical locations."

Jake said, "So you're implying the movement really has wider applications, right?" and Bob gave a quick dip of her chin, took a bite of oatmeal, and carried on.

"The ideal location for a single-, or multi-unit residence, or a commercial building for that matter, would have enough area to allow a longer east–west orientation, facing as close to south as possible. It is important to have a wide-open, solar horizon, either controlled by the property owner or preserved by an easement."

Abbey asked her if a solar horizon meant the whole, unobstructed circuit of the sun from morning to night and during all seasons of the year, and was told "yes."

"From this simplified sense of a rectangle, the intuitive roof design would be a ridge with a parallel east–west axis—"

"Which gives you a wide, south-facing pitch of roof for solar panels," Alice said. "Combine that with a standing seam metal roof and you gotta perfect place to mount the array on racks designed for these roofs."

"And wildfire protection," Abbey added. "If you avoid having trees too close anywhere around, then that's a great first step—especially if you've built into what Jake has explained is called the 'wildland-urban interface.' Keeping your gutters clean is just another bonus," she said, signaling with another sip of juice that she was done speaking.

Alice commented, "Abbey, you're avoiding coffee for the pregnancy then?"

Her reply was that it wasn't that hard to give up.

"So . . . windows," Bob said as she put another half-piece of bacon on her plate. "The solar-home movement began years ago with a sort of 'more is better' ethic, and often early practitioners placed too many windows facing the sun."

"Turned their structures into solar cookers, really," Jake said, laughing.

"Exactly," Bob said spiritedly. "But really good windows should be used, just more sparingly. Triple-pane windows filled with an inert gas like argon, if possible. And shading is hugely important—a calculated, south-facing overhang welcomes the winter sunlight, but almost excludes it at the height of summer.

"People know west windows without external shade can really heat up a structure. What they fail to understand is that east windows allow in the same amount of energy, just at a cooler time of the day. So minimize these windows too. And you can use trellises or fences at both ends to give them some direct shade. Grow vining plants on these and keep them watered."

Jake considered all this, saying, "So, what we have is a hypothetical rectangle with the long wall and the parallel ridge of the roof facing as close to true south as achievable. No trees near the house. Triple-pane windows, but not too many. Which is good, because they're expensive.

"What's next?"

Bob was ready. "The Passivhaus detailing is critical. The house we're going to start building in the spring will be a three-bedroom, two-and-a-half-bath place for us and the two kids, with our bedroom on the second floor and really expansive views up and down the valley. Just shy of nineteen hundred square feet.

"We bought the property with a good well two years ago, with certified water tests by the public health department and a certified output of eight gallons of water per minute. This year we had the gravel driveway improved and extended, a septic system installed, and a pair of three-hundred-foot trenches excavated, three feet wide and five feet deep. The installer put two runs of heat exchange tubing in each trench, for a total of 2,400 feet underground. Finally, a separate excavation for burying the electric and cable services. Then, all the trenches were backfilled, ready for spring."

"What are the two long trenches for?" Abbey said.

"A beautiful setup for a ground-source heat pump is what I'm thinking," Jake replied.

"Both space heating and cooling, as well as domestic hot water," Alice confirmed.

Abbey got up to get some more oatmeal, adding raisins and a pinch of brown sugar.

"Okay," Bob said, "time to hit the wall. These houses are super-insulated. We chose a hybrid wall of insulated panels on the outside, stick-framed on the inside. So, a pair of six-inch walls sandwiched together and attached at the top and bottom plates. The outer layer is composed of six-inch-thick SIPs—'structural insulated panels'—in which graphite is included for structural strength, even though they have a half-inch OSB on both sides."

"What's OSB again?" Alice said. "Oriented . . . something . . . what?"

"Oriented strand board, remember?" Bob said. "An engineered product where wood shards are pressed together and glued into position. Sort of like plywood on both sides, making it one solid unit. What is fantastic is that on a computer, all these wall sections are designed and robotically manufactured, then openings for doors and windows are cut out before stacks of them are delivered to the construction site on flatbed trailers.

"With SIPs, some adjacent edges will have the insulated core, or alternately, the two OSB layers standing proud. On the bottom, the OSB layers are forced down over the wooden bottom plate with lots of adhesive applied, then nail guns go to work from both sides penetrating into the plate."

"Same sort of joint vertically, with a hole drilled through but not too close to the edge. Come-alongs are attached to a bolt in the hole. Lots of, how do you say, elbow grease is used to get all of them seated into position and secured with adhesive and nails."

"Lots of adhesive. Lots of nails," Jake said, putting some of the last dishes in the rack. The others were nursing their final cup of juice or coffee.

"I observed the construction of some of the projects in Sweden that my parents designed. An intricate dance of people and tools. But as tough as it is, a good crew can put up five or seven panels a day, per team.

"The top sections forming the roof or ceiling can be twelve inches thick and hefty. A couple of days of work with a crane are required to maneuver them into position resting on wooden trusses two feet on center." Bob held up her arms in a large, inverted V-shape. "But the wall sections can be put into place with three or four workers.

"The outside half is the hard part. The inside half of the wall is typical six-inch wood framing, which is conventional construction work, with the top and bottom plates doubled up for strength and directly sistered onto the inside of the SIPs. This part is attached to the foundation with J-bolts, the kind you see sticking up at construction sites once the concrete of the perimeter footing is poured. Usually with hard, orange rubber caps on the J-bolts for safety until the wall is mounted in place.

"The beauty of a hybrid wall is that all the plumbing and electrical runs can be installed here with all the open space once the building is dried in—"

Abbey cocked her head. "Dried in?"

Bob knew she'd gotten ahead of herself. "Yes, for certain. When the whole outside part, or envelope, of the structure is complete. A semipermeable membrane is attached to the outside of the walls—you've seen the common trade names and logos there—with a similar treatment on the three-quarter-inch plywood sheathing of the roof deck. Usually, all the wood has been through multiple rainstorms by this point, but can finally start to 'dry in.'" Bob held her arms out with hands up to demonstrate.

"Sadly, from my point of view, once all the plumbing is roughed in and all the electrical junction boxes supplied with wiring runs, then an internal membrane goes up everywhere on the interior walls. The stick-framed half of the wall is filled chock-full with blown-in insulation. You'll see beautiful, finished wood displayed around doors and windows, but all the rough framing gets buried, never to be seen again."

Jake saw Bob give a big sigh. "But what's the result of all this work? How good are these walls?" He drummed his fingers once, just a quick tapping run.

"Insulation values are way above code. I mean to say, resistance to heat flow for exterior walls is usually required at R-13 to R-23 in the building code. But these walls are over R-50, and the ceiling or roof, depending on design, tops out at over R-80."

"Sounds extraordinary," Abbey said. "Though I'm not sure what it all means."

"It is extraordinary," Bob said, "but I left out the best part. These houses are really tight in terms of infiltration of air and migration of heat in or out of the building envelope. How do they test that? They will do blower-door testing at several steps in construction after all the windows and exterior doors are installed, before the inner wall is filled with insulation. All the various vents and pipes are caulked and taped and temporarily capped off, then a single test door is installed with a central fan trying to pull air sneaking in anywhere through the envelope, even openings as small as the head of a pin.

"The workers sometimes carefully carry candles or other smokers, inspecting for tiny puffs of air or smoke. The Passivhaus consultants local to our area will inspect construction techniques throughout, but these examinations also rely heavily on verification. The architects and builders put together an electronic file of all the specs for the house, and the Passivhaus decision will be anxiously awaited, no doubt."

Jake used a bread knife to work around the muffins, then extricated each and placed them on another platter for the table, very well received, then put two on Abbey's plate at the counter for them to share.

"These are wonderful, thanks so very much," Alice said.

Bob echoed her compliments, then continued after chewing her first bite. "You might wonder what the situation is in terms of electricity supply in Bryson City," she said. "Residential electricity rates are highest in January, but the highest average bill is in November. Our electric rates average 10.28 cents per kilowatt-hour, substantially less than the average national electricity price of 15.5 cents as of 2023."

"Who's your utility in Bryson City, again?" Jake said.

"Duke Energy," Alice replied.

"Ah, that's what I thought," he said, his eyes narrowing. "Duke Energy is the nation's largest investor-owned utility, as you no doubt know. They've got a mix of almost every kind of generation. Gas and nuclear are more than one-third each, and they still have a lot of coal burners, a pretty dirty combination all around. Renewables are less than two percent."

Alice seemed taken aback. "Boy, you know your stuff—"

"This is what I research all the time. I'm sorry if I sound like I'm dissing your utility, but I have strong criticisms about this company."

"Don't worry," Bob said. "Clearly, we share your feelings about energy in general. Besides the dirty energy portfolio, they're also fighting tooth and nail against giving up their fossil fuels. I think they want to keep living in the early twentieth century. In at least five of the states where they have customers, they have aggressively lobbied for weaker net-metering programs and for higher monthly connection charges for solar systems."

"Delighted to hear you share my feelings. But you should read an annual report to get a sense of their propaganda."

"I have, Jake, I have. It is part of my responsibility as city manager to be aware of these issues."

Whoops, he thought, *foot-in-mouth disease again.* "So, they do still allow net metering? No quotas for the proportion of solar hookups? No jacking up connection fees?" His fingers still felt restless.

"We are planning to . . . push the plow down the field—how do you say it?"

They all laughed with her, lightening the mood a bit.

Alice, practically bouncing in her chair, said, "We should get our certificate of occupancy by the end of next year, which means we'll get the federal tax credits for the ground-source heat pump, solar panels, inverter, and batteries."

"And how excited are the kids? And how much battery backup?" Abbey said.

Jake was proud of her for asking about energy storage.

Bob took the reins. "They are both thrilled. We are designing a separate electrical subpanel just for the most critical circuits in case of grid failure. Refrigerator, freezer, kitchen counter for coffee maker," she said as they all smiled in agreement, "two separate computer stations, entertainment center."

"Lights in the kitchen and the two full bathrooms," Alice added, as she poured another glass of juice.

"Septic pump and well pump, of course," Bob said. "Almost easier to list what we leave out: big appliances like the microwave, oven, and induction cooktop. Just too much of a draw. Garage at each end of the house with greedy level-two chargers, but the single vehicle we are planning for now will be a pure electric car with enough room for the four of us . . ."

"And kids get big fast, I hear," Abbey said.

"Tell me about it," Alice said happily as they grinned at each other.

Bob continued, "And in an outage, we think any car will easily maintain enough charge for a week of driving locally, so that's not really a concern."

"And you could always drive away from the outage area to find a place to charge, right? Are you thinking of using the vehicle or house batteries as backup for the utility itself? In demand response?" Jake queried. *Hope so.*

"As you might guess, Duke Energy is not open to this yet. But management, politics, policies, and the times all change. We'll see," Bob said, leaning back and sipping her coffee.

Abbey sat up straighter. "I don't know if Jake told you last night, but we're just caretakers for my parents' place here, running the Inn 'til they get back from overseas."

The two other women just shook their heads.

"We'd like to build such a house ourselves. It may even be in this part of Virginia, as we both work in the capital. Jake just started teaching at

GW, while I've joined the practice of another infectious disease specialist, and he and I consult at several hospitals in the city."

"I'm thinking we should get the right vehicle first—pure electric certainly—partly because the federal tax credit's still available, partly because maintenance visits are few and far between, and partly because this would shield us from the crazy swings in the cost of gasoline," Jake said, tapping his fingers again.

"So, time for the nitty-gritty. How do you justify paying for all the special features in a Passive House?" Bob looked over at Alice, who gave her a quick lesson in slang.

"The first part's simple, and we just discussed it. An EV over its life cycle costs less than a fossil fuel car, so every time the price of gas goes up, we're just paying it off faster. Not to mention," Jake threw in, "you'll be taking away demand from the oil and refining industries and taking advantage of a domestic source of energy, or a universal source—sunlight." He slowly shook his head then raised both eyebrows.

"The next part is almost invisible," Bob said. "Unlike almost all folks in Swain County, we will not be depending on propane or methane gas or dirty heating oil. Our appliances will be superefficient and the house, all-electric. I've talked with a lot of the people in Bryson City, and some of them have gas bills of two thousand dollars or more a year for cooking, heating, and hot water."

Jake put his elbows on the counter. "So, you accomplish another destruction of demand for the fossil fuel industries—good work," he said, slowly shaking his head, eyebrows raised.

"Since you ask," Bob said, "the cost of the Passive House features at this point are in the range of custom house building. Any surplus is usually for the architectural design and fanatical detailing necessary by construction teams, which is paid for in energy savings in at most, several years."

"If you factor in the systems you don't need, like a furnace and air conditioner, then you might not have any increase in cost at all," Jake said.

Alice pursed her lips. "I said to Bob, if we rightsize the solar system, all of our transportation will be paid for."

Jake asked if either one of them knew how much the irreducible monthly connection fees for the utility would set them back.

"I get asked a lot of questions about the Low Income Home Energy Assistance Program run by Duke Energy," Bob said. "I scrutinize many utility bills, though we can't offer much of a break sometimes. It varies by location, but seven dollars and fifty cents a month for us. About ninety dollars a year—"

Jake whistled. "Are you kidding me? You're going to live in an all-electric home, and your entire energy bill and all your local transportation is only going to set you back ninety bucks a year? Sign us up; I'd love to do that, right, Abbey?" She couldn't stop smiling, either.

"Holy Toledo, what'll they think of next?"

Bob looked at Alice again. "What is this Toledo he is talking about?"

Alice doubled over, laughing. "It . . . it's in . . . Ohio," she said, slapping her knee. Jake and Abbey let out peals of laughter, until Alice finally took pity and tried to explain to Bob.

Breakfast, clearly, was over.

Jake was ready to get his gear on and knock off some shoveling while these two wonderful people could still get out to see the sights. And build back better this year, and next, with their innovative, climate-change-friendly house.

TWENTY-FOUR

Time Is Short

The six stages of climate denial are: It's not real. It's not us. It's not that bad.
It's too expensive to fix. Aha, here's a great solution (that actually does nothing).
And—oh no! Now it's too late. You really should have warned us earlier.
—Katharine Hayhoe

Jake and a small group of faculty were down front conversing about something that must have been comical, because they kept shaking their heads and laughing. Max Baerbock and Mikhail Ligachev were the speakers for tonight, but Addie Higgenbotham and other members of the geology department were attending the final event of the year as well. Finally, the hour arrived, and Jake turned to face the audience in the auditorium, held his hands up, and said, "Folks, really, you can't make this stuff up. We're going to be talking tonight about all the potential changes we face with global climate weirding, including flooding and droughts and other changes in the hydrologic cycle.

"But, as some of you are no doubt already aware, a large water main break occurred late this afternoon over on Twenty-Second Street Northwest, and if you tried to get here tonight from that direction, you were obviously detoured away from that particular corridor. The District of Columbia Water and Sewer Authority, which most of us know as DC Water, has notified campus authorities that a large pipe broke. They estimate it'll take at least ten to twelve hours to get that repaired. Fortunately, no rain is forecast for tonight to compound the problem.

"But you could float a canoe or kayak over there right now. At least parking is apparently free.

"Unfortunately, the Science and Engineering building is actually affected: We have minimal water pressure, with nonfunctioning water fountains and sinks. We cannot even guarantee that the toilets work, so you might try checking by flushing first. This may sound like a breakdown of the thin veneer of civilization, but if it's only urine, then you might not need to flush at all.

"Finally, water purity is apparently questionable until we've done testing. So no, it's not safe to drink the water. People wonder how flooding and drought can be juxtaposed, and tonight happens to be a practical demonstration of that very circumstance.

"Everybody I assume knows Murphy's Law, namely 'whatever can go wrong will go wrong.' It turns out Murphy had a drinking buddy by the last name of O'Connor, who claimed one night in a pub, 'Murphy, me friend, whatever cannae go wrong will still turn to shite.'" After a moment, folks appeared to get the joke and guffawed.

"So there you have it, O'Connor's Corollary to Murphy's Law. In terms of historical veracity, I believe Murphy was actually an aerospace engineer from southern California, but the Irish pub sounds more credible, somehow.

"This water pipe snafu may seem like something out of the theater of the absurd, but in reality, I just checked, and North America experiences about 850 water main breaks daily,[43] with metal pipe corrosion being the typical culprit. Errant backhoes don't help, either. You don't hear about them unless they're local, because the press only reports on about one percent of these events.

"Internationally, according to the World Health Organization, one third of the global population does not have access to toilets.[44] Water systems are thus often contaminated—both surface and subsurface—and water-associated diseases are rampant. Nearly a billion people must relieve themselves outdoors. India is the worst off, with 640 million people engaging in open defecation.

"But climate change will challenge sanitation in this country, as well. In Florida, 1.6 million households rely on septic systems.[45] With sea level rise directly and with saltwater intrusion below the lenses of fresh water, the nonsaturated volume of soil for these systems will be compromised. Once these systems fail—and most assuredly they will—then aside from measures such as composting toilets, much more expensive, extensive, and elevated municipal wastewater treatment is the only alternative.

"Okay, so our first speaker tonight is going to be Professor Max Baerbock from the physics department. He'll be followed by Professor Mikhail Ligachev from the geology department, brought back for a return engagement. Professor?" Jake motioned to Max—middle-aged, tall, and balding—who stepped forward. He sported a bow tie with flowers that somehow polished off the picture nicely.

"Thanks, Jake. I'm glad to be here, even under such inauspicious circumstances. Mikhail has already explained Milankovitch cycles; the top of the atmosphere, or TOA; and other concepts of the climate system. You now understand that the planet is warming, not because of any increase in sunlight, but rather because of greater trapping of the Sun's heat by trace atmospheric gases. You know that the three principal gases trapping heat are carbon dioxide, methane, and nitrous oxide—aside from water vapor, which is transient and variable and outside of our control, given that we live in a 'water world.'"

"Scientists quantify the trapping of heat by thinking of it as additional watts—of heat—per square meter at a hundred kilometers above the Earth's surface. This is the energy of infrared radiation attempting to escape the planet, which is instead absorbed by an outer electron of a greenhouse gas at all levels of the atmosphere. That excited electron expeditiously drops back down to its ground state, and about half of the reemitted infrared heads toward the surface while the other half is sent into space. Completely random in each individual case."

He stepped over to the whiteboard and wrote:

$$Wm^{-2}$$

"This is the shorthand expression for watts per square meter. It represents that fraction of the extra infrared returned to heat the planet by the more plentiful insulating greenhouse gas molecules. And you may recall from Dr. Ligachev's lecture several weeks ago that heat in the form of infrared or heat radiation is constantly attempting to flee the Earth in all directions through the TOA, which calculates to 526 trillion square meters. One of those big numbers nobody tries to memorize, honestly.

"So even a small amount of heat-trapping adds up to a huge number. The current value is only about 3.2 Wm^{-2}, which is hardly the rating of a small LED light bulb. But multiply that by the huge number and you get the additional heat that the land surface—and initially the photic zone of the ocean, down to about two hundred meters—must absorb and accommodate.

"One important concept is the lag time between the augmentation of heat-trapping gases in the atmosphere and the secondary accumulation of heat. Keep in mind that if we stopped all combustion of fossil fuels today, the temperature of the atmosphere and ocean would continue rising noticeably for only three to five years. But it would take much longer to reach equilibrium." Max paused for a moment and absentmindedly loosened his bow tie, then wrote:

Climate Sensitivity

"Are you familiar with the phrase 'climate sensitivity'? Or the concept of 'radiative forcing'? I am seeing some people shaking their heads, so let's nail down, if we can, the meaning of these central concepts.

"Beginning in the 1890s, Svante Arrhenius in Sweden began asking a simple question. Looking at preindustrial times prior to 1750, and extrapolating into the present and future, he wondered how much warming would result from a doubling of atmospheric CO_2. By warming, I'm specifically referring to a change in global mean surface temperature, or GMST, including the ocean and all other bodies of water.

"A multitude of studies using climate models since the 1970s may have been superseded by a controversial warming estimate by Hansen, et. al. in the range of 3.6 to 6 degrees Celsius from a doubling of atmospheric CO2. In the late 1700s, the atmosphere contained about 280 ppm of CO2, but we are now up to over 420 ppm, an increase close to fifty percent. By the end of the century, we could reach 600 ppm or more.

"The most extensive review of climate sensitivity was accomplished over the last seven years by a consortium of over eight researchers in fifteen countries, looking at the last sixty-six million years—since that asteroid hit. This new model, published in the journal *Science*, argues that a doubling of CO_2 would warm the planet by a whopping 5 to eight 8 Celsius—albeit over thousands of years."[46]

Jake kept up his furious pace of notetaking while noting the tapping of fingers on keyboards.

"By the end of the last decade, we reached a 1.15-degree-Celsius increase in global mean surface temperature. I'd like to think I'm joking when I say it would take another coronavirus pandemic and subsequent worldwide recession to have any hope of not exceeding 1.5 degrees Celsius, but by now it's clear that we need not just stabilize but claw back the levels of heat-trapping gases.

"At the very heart of climate science lies this issue of the overall climate's sensitivity to an increase of atmospheric carbon dioxide. What was specifically selected as a metric is the doubling of carbon dioxide from the preindustrial baseline of 280 parts per million—therefore, seeing that figure go up to 560 parts per million. Understand that this target wasn't arbitrarily chosen, but had been in use for over a century. I should note that other heat-trapping gases are often rolled in as so-called 'CO_2-equivalents.'" Max put one hand on the lectern for a moment.

"The Intergovernmental Panel on Climate Change was formed under the auspices of the United Nations and World Meteorological Organization. Their first report was in 1990, and every six years since then, they have produced a set of documents, the embodiment of the work of hundreds of volunteer scientists in each iteration. The 'I' in IPCC is often mistaken for 'International,' and in fact, many countries contribute to these publications. But 'I' for 'Intergovernmental' emphasizes that nonscientist representatives of every government—rather than the scientists you've been hearing about—may review, modify, or water down the conclusions."

He turned to run his gaze across the whole audience. "In case you've been assuming rising temperatures are solely a straight-line extrapolation based on heat-trapping gases produced by humans, let me swiftly disabuse you of that notion. The Earth climate system is very complex. Our fear is that exceeding 1.5 degrees Celsius—even a rise of 1.0 degrees Celsius for some of these critical factors—could begin triggering multiple climate tipping points. These are thresholds beyond which a change in a part of the climate system becomes self-perpetuating, with abrupt changes due to tumbling dominoes.

"I'm a natural optimist, but rationally we recognize that the world is heading toward a warming rate of roughly two to three degrees Celsius. Only if all net-zero pledges and nationally determined contributions are implemented could we possibly stave off reaching a rise of two degrees Celsius. A recent keystone paper published in the journal *Science* synthesized 'paleoclimate, observational, and model-based studies, [providing] a revised shortlist of nine global "core" tipping elements and seven regional "impact" tipping elements and their temperature thresholds.'47

"Obviously, we don't have time tonight to discuss the elements of all global and regional temperature tipping points, so I've selected three of them to examine in more detail, denoted by asterisks. However, this slide comprises the whole collection."

Climate Tipping Points

- Arctic winter sea ice [AWSI]
- Barents Sea winter ice [BARI]
- Greenland ice sheet [GIS]
- West Antarctic ice sheet [WAIS]**
- East Antarctic subglacial basins [EASB]
- East Antarctic ice sheet [EAIS]
- Labrador–Irminger Sea convection [LABC]
- Atlantic Meridional Overturning Circulation [AMOC]
- El Niño-Southern Oscillation [ENSO]
- Indian summer monsoon
- Sahara/Sahel + West African monsoon
- Amazon rainforest (AMAZ)**
- boreal forests [BORF]
- boreal permafrost
- mountain glaciers [GLCR]
- low-latitude coral reefs [REEF]**

"Before I discuss these, you should know the other elements that didn't make the cut. But over time, continued examination may lead to the deletion and replacement of some of the current representatives. Such proposed concerns include but are not limited to southwest North America, the northern polar jet stream, and the Congo rainforest."

Jake and others were scribbling and typing at frantic rates. A couple of folks even raised their phones to take a quick snap of the slide.

Max gave them a chance to finish while he put up his next section title:

West Antarctic Ice Sheet

"Let's start with the West Antarctic ice sheet, large parts of which are grounded below sea level. Remember that grounding lines are the point at which glaciers and ice shelves start to float like huge ice cubes. If those grounding lines retreat far enough to where the bedrock slopes toward the middle of the continent, this may produce 'marine ice sheet instability,' causing self-sustaining and unstoppable retreat. This would occur over centuries, but there is nothing conceivably that humanity could do that would rebuild this big ice which took 34 million years to create.

"The best-known glacier, Thwaites, is 120 kilometers, or 80 miles, wide, and only about 30 kilometers, or 19 miles, away from a last-ditch subglacial ridge, retreating now a kilometer per year. If the glacier melts and ends up floating above this rearguard obstruction, then relatively warm subglacial water can race forward many kilometers farther under the ice. This glacial system is a bulwark, a beaver dam, holding back a lot more ice sheet behind it.

"This phenomenon, known as 'grounding line retreat,' is not necessarily fixed and linear. Arguably, eventual collapse may already be inevitable. Researchers are confident that West Antrctic ice sheet deterioration is a core global tipping element, incorporating a best-estimate threshold of 1.5 degrees Celsius, but with a range of 1 to 3 degrees Celsius, which overlaps the current condition. They have medium confidence that the timescale to complete melting would occur in two thousand years, with a range of five hundred to thirteen thousand years. Associated sea level rise was recently recalculated to 4.2 meters, or about 14 feet."

Jake looked down at what he had just written and thought about each of the cities that would be turned into Atlantis forever. *And all of the beaches I've ever known.*

Max wrote again:

Amazon Rainforest

"Our second tipping point is the Amazonian rainforest, a biologically rich forest biome storing some 150 to 200 billion tonnes, or gigatonnes, of carbon. Since the 1970s—with an acceleration since 2019—seventeen percent of the rainforest has been lost to deforestation. Even in areas that haven't been wholesale deforested, the carbon sink has declined since the 1990s, driven by climate-change-induced desiccation and especially human-induced degradation in Brazil's east and south. Almost inconceivable to me is that the Amazon is becoming a net carbon source—no longer a sink.

"Widespread Amazonian dieback was originally projected as a source at three to four degrees Celsius of warming, or an estimated forty percent deforestation. But uncertain interactions with factors such as drought or wildfires could conceivably lower the threshold to only twenty to twenty-five percent of the whole Amazon area, not far from the current proportion of loss. Our best estimates for seeing the Amazonian rainforest convert to a relatively barren savannah are a threshold of about 3.5 degrees Celsius with a range of two to six degrees and timescales of about a century with a range of fifty to two hundred years, leading to emissions of about thirty billion metric tons of carbon."

Jake realized Max wasn't going to make any reference to autocratic government or human rights abuses or mention the reassuring fact that Brazilian president Jair Bolsonaro failed to get reelected.

"Third and last for tonight's purposes is coral reefs." Max took a moment to write that on the board.

Low-Latitude Coral Reefs

"Tropical and subtropical reefs are suffering from direct damage, sedimentation, overfishing, acidification, and especially, marine heat waves. Warming water causes corals to irreversibly expel their symbiotic algae, resulting in bleaching and coral death. Coral collapse would remove a diverse biosystem, home to half the ocean's nurseries of fish, with knock-on effects on wider marine food webs, carbon and nutrient recycling, and ways of life for millions of people. I've skin-dived and scuba dived on bleached coral, and the reefs are a shocking, ghostly whitish-gray. This is often followed over time by dark algal overgrowth.

"The IPCC projects seventy to ninety percent reef loss at 1.5 degrees Celsius, with essentially a total loss of the reef if the temperature increased two degrees. While this is a regional—not global—tipping point, their best estimates are a threshold of 1.5 degrees Celsius with a range of one to two degrees and timescales of about ten years, but fortunately, negligible feedback on global mean surface temperatures.

"Our most urgent concern must be the potential activation of five of these dangerous tipping points, linked to the current global mean surface temperature rise of 1.2 degrees Celsius. The collapse of Greenland's ice cap, the halt of the conveyor belt of the Atlantic Meridional Overturning Circulation in the North Atlantic, the early, abrupt melting of carbon-rich permafrost, the collapse of Western Antarctica's ice sheet, and boreal, or taiga, wildfire destruction are the five elements posing the most profound risks. Our concern must in part be based on a potential domino effect, which may accelerate these degradations."

Max took a deep breath and looked at the slide he had just discussed. "Let's move on."

Jake continued taking notes, as did all. But he also thought of his own skin-diving and scuba diving experiences, the bleached coral in Hawaii and the Caribbean. *Gotta save coral, dammit.*

"The Fifth Assessment Report, or AR5, from 2014 established a new set of scenarios as a framework—called 'Representative Concentration Pathways.' They did not choose the four scenarios of climate change arbitrarily, but rather as a basis for assessing the risks of crossing identifiable thresholds of physical change and the consequent effects on biological and human systems—as set down in the IPCC technical summary."[48]

"Keep in mind that if the many climate modeling teams around the world each used individualized metrics—particularly with different assumptions about baselines and eras—then it would be virtually *impossible* to compare and contrast their results.

"The other rationale for these RCP scenarios is that the supercomputers required to make these calculations—both retrograde to test these models and antegrade for future prediction—are expensive, and in short supply but great demand.

"The number associated with each scenario is the approximate increment in net radiative forcing in Wm^{-2}—the atmosphere's additional heat-trapping—by the end of the century. And this slide summarizes the four defined pathways."

Representative Concentration Pathways

Pathway	Radiative Forcing	Estimated CO_2
RCP 2.6	2.6 Wm^{-2}	421
RCP 4.5	4.5 Wm^{-2}	538
RCP 6.0	6.0 Wm^{-2}	670
RCP 8.5	8.5 Wm^{-2}	936

Scenarios assume no major volcanic eruptions or changes in solar flux

---

Rachel raised her hand and spoke up. "Professor Baerbock, Pathway 2.6 assumes a radiative forcing of 2.6 Wm^{-2}, but we are already past 3.15 Wm^{-2}.[49] So how does that pathway make sense as a future scenario?"

"Excellent question. This is the most, most hopeful scenario—aspirational in the sense that this pathway assumes we can not only stabilize the CO_2 level but can drive it back down rapidly. It relies on adopting both natural systems, such as reforestation or biochar in soil—which is biological material that decomposes in the presence of little or no oxygen—and quite probably mechanical devices to extract CO_2 from the air and sequester it.

"This latter idea is geoengineering, which I would never advocate unless we were irretrievably pressed to the wall. But with these approaches combined, we could reduce the excess radiative forcing back down to 2.6 Wm^{-2} and get back the planet we had at the beginning of this century. Make sense?"

Rachel replied, "I think we're already all on board."

He chuckled and went on. "Look at this next slide. This is a graphical display of the four pathways to the future. The gray column in the middle represents the whole twenty-first century, where we are now."

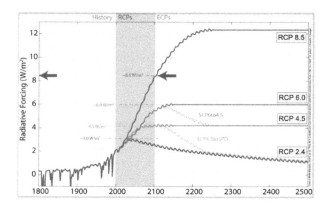

Figure 10: Potsdam Institute for Climate Impact Research, Global Anthropogenic Radiative Forcing for the high RCP8.5, the medium-high RCP6, the medium-low RCP4.5 and the low RCP3-PD, central estimates.

"The right edge of the column shows where each of these pathways will land us in the year 2100. But really, this just tells us each pathway's prediction of how the climate will change in the short term, right? After all, we're talking about less than eight decades to the end of the century. After 2100, these scenarios become what we call 'Extended Concentration Pathways,' which are more speculative at this point.

"As an example, you can see the Pathway 8.5 trajectory crosses right at 8.5—hence the name—in the ordinate of this graph on the right side. See those dark arrows marking that level? Each of those four curves levels off over the next couple of centuries—the 'equilibrium climate sensitivity'— although Pathway 2.6 is the beguiling exception as discussed. That leveling off would demand significant personal and political change—a true sea change. Otherwise, we'll see a change in the sea . . . sorry, couldn't resist that one."

A student high in the back row put her hand down and spoke up. "Could you please explain what radiative forcings are again?[50] It's not exactly intuitive."

Max walked over to face the student. "Of course. A radiative forcing is a 'measure of the effect any factor has in altering the net balance of incoming and outgoing energy in the Earth-atmosphere system, measured in watts per square meter.'" He pointed again to the Wm^2 he had written earlier on the whiteboard.

"Remember, this measure is the heat equivalent, not the actual watts of electricity. I hope I made that explicit before. Components of these reactive forcings can be natural, such as volcanic eruptions and orbital variations, as well as human-caused or anthropogenic. This next bar chart outlines some of the major forcings, which can have a potent effect on warming the planet."

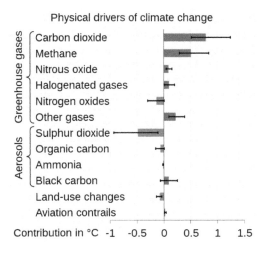

Figure 11: Image by Eric Fisk, licensed under CC BY-SA 4.0

"Bars to the right correspond to changes in net warming since 1750, while bars to the left side show net cooling. The top six comprise the collection of greenhouse gases, which represent our predominant human influence—or forcing—on climate."

Brian asked why this chart did not include previously described climate control knobs, like the variations in Earth's orbit and the weathering of rocks.

Max distinguished in his reply that the climate control knobs were omitted because of their insusceptibility to human manipulation, rather than due to the slow pace of these forcings.

Brian nodded and gave him a thumbs-up.

Max continued, returning to pacing. "Couple of other caveats worth mentioning. Halogens—like the infamous chlorofluorocarbons or CFCs—have both a positive forcing as greenhouse gases and a negative forcing in their destruction of stratospheric ozone. Volcanic eruptions have been omitted since IPCC AR5 because their wild or unpredictable fluctuations in radiative forcing make comparison to other forcing agents well-nigh impossible.

"Okay. After a short break, Mikhail's going to take over and talk to us about carbon budgets, sea level rise, and international agreements that actually demonstrate how the world can address climate disruption."

Rachel whispered to Brian, "Next comes the encore to *Götterdämmerung*," and he cracked a smile and shook his head. They saw Jake going outside with a phone held to his ear.

TWENTY-FIVE

Worm of Suspicion

Suspicion is a virtue as long as its object is the public good, and as long as it stays within proper bounds. . . . Guard with jealous attention the public liberty. Suspect every one who approaches that jewel.
—Patrick Henry

Jake opened the door and walked down the hallway. "Give me a moment, there are a lot of people here, Emmanuelle. Don't worry, we're on a break for a couple of minutes.

"Okay, Emmanuelle . . . didn't really expect a call from you. Assuming here you're not trying to set up another reservation. So . . . what's up?"

"If this is not a good time, I could call you back later."

"No . . . I have about ten minutes," Jake said, getting more curious.

"I'll get right to the point, then. Sometime, long after we stayed at Dragonfly Inn, you sent me an email about somebody you'd met who seemed fishy somehow. I wanted to get back to you, but not by email, which is notoriously leaky. I decided instead to call you directly. As a courtesy, since I'm not required in my position to do so, I'll simply say that I'm recording this call. Can you tell me more about him?"

"I've never had anyone say that to me before," he said, chuckling to himself. He scratched his head. "At least—not that I can recall."

"I would ask you not to repeat that part or any part of this conversation to anyone."

Here, he was really brought up short. "This isn't something subject to the Official Secrets Act, is it?" He felt odd just to be asking.

"No, Jake, that would be the UK, not the US."

He licked his lips, then remembered the time pressure. "I met this weird, kinda suspicious guy wearing sunglasses during a break in the lecture back in November. These two-hour seminars are so long that we split them in half and give people a chance to get up and move about a bit.

"This guy was someone I'd never seen before and haven't seen since," Jake said. "I am now inferring that this person only seemed to be interested in nuclear power while Enoch Apfelbaum—whom you said you know—was addressing that topic."

"What'd this guy say to you, Jake?"

He thought she sounded calm but intent. *Bet she's a good interrogator.*

"I've gotta say he seemed like an odd fish. Wore sunglasses the whole time, even inside. Lanky, like an athlete, really pale. He never actually looked right at me. Intense, low voice—"

"If you were to guess at an accent, what would you say?"

"He comes from somewhere on the East Coast, except not Virginia."

"No trace of a foreign accent?"

"I didn't think so at the time."

"How well-spoken was he? Would you guess he had a college background?"

"I think he asked me where I went to school, and I told him, but then he was evasive when I asked him about his schools. Just said something about several places—not around here. But he certainly didn't sound unintelligent."

"Was he wearing a hat?"

"Yes, like a baseball cap, but with no team logo on it. That and the glasses made it hard to see his face."

"What would you guess would be his height and age?"

"Maybe five eleven, an inch taller than me, early thirties."

"No facial hair?"

"Just sideburns, a little longer than usual."

"Anything noteworthy about his clothes? New? Old?"

"Blue jeans, no fancy label. Shirt could have even been secondhand. He was wearing running shoes—boy, those looked new. Nike, I think."

"No obvious scars, tattoos, piercings, jewelry, watch, rings?"

"Not that I remember. I'm thinking maybe I'm not as good an observer as I thought I might be in a serious situation."

"No, Jake. You're doing just fine. That's part of why I wanted to respond to your reaching out to us. After our lively conversation at the Inn, I sensed you were a thoughtful guy. Now, this is important. What did he say that made your ears perk up?"

"Once he found out I shared Enoch's reservations about nuclear power, he said something like, 'But don't you think we should be doing something about that? Doing it right now?' Then he went back inside without even waiting for an answer and returned to the top row of the auditorium. At the end of the lecture, I got to talking with Max Baerbock and I lost track of the guy. Didn't even see him leave."

"Jake, you should know that the FBI gets thousands of tips a year, and I will say this sounds like a soft one. But you have a gut sense about this guy, don't you? Or you did, once you thought about the encounter afterward?"

He was listening to her voice carefully and understood this marked a decision point. "I don't want to be written off as alarmist, but yeah, I am kind of worried about this guy."

"I know we don't have much time, but are these lectures recorded?"

"Regular classes are not. But these are."

She was quiet for a few moments. "Do you think the videographers recorded a bit before and after the seminar as well as during the intermission?"

"Don't know, but you could check with—"

"That won't be necessary, Jake, for obvious reasons."

I can guess what those reasons are, he thought.

"Let me just remind you not to share anything about this call, even with Abbey. Which means you can't even say hi to her for me."

He was going to have to think about that implication.

"We'll get back to you if we need any further help. If this doesn't pan out, which is extremely likely, I'll never mention this to you again."

"Okay." And with that, she was gone, and he headed back inside.

Now I'm beginning to wonder how many people were really on the call.

TWENTY-SIX

And the Water Is Rising

By the time Jake had returned to his seat in the front row, Professor Ligachev had taken center stage and begun speaking. "Again, it is good to see so many people—especially members of the general public—interested in what used to be an arcane area of science, namely the study of the climate system."

Jake was still shaking his head, unable to dismiss his intuition something was seriously wrong with the "sunglasses guy." *Really out of kilter.*

"Let's start with the concept of a carbon budget," Mikhail began. "The Kyoto Protocol incorporated the understanding that we should strive to keep the Earth's surface temperature rise under 2 degrees Celsius, and preferably under 1.5 degrees. The risks of not doing so would otherwise consist of triggering positive feedbacks, such as the disintegration of permafrost, the melting of methane clathrates (crystalized methane), decreased albedo (sunlight-reflective capacity) of melted ice and snow, and the conversion of tropical and temperate rainforest to biologically impoverished scrubland. The next slide references an article in *Science* that calculated the remaining amount of carbon we could release into the atmosphere and still stay under the 1.5-degree limit."

Short-Term Carbon Sinks Overflowing
- proposed limit of 420 Gt, or gigatonnes, additional after 2018
- 2022 reached 36.6 Gt from GHG, marginally less than pre-pandemic record 2019
- equivalent to 42 Gt with land-use changes included
- clean energy tech such as EVs + heat pumps prevented 0.55 Gt

"It should be apparent that 2022 was not only a near-record year for carbon being added to the atmosphere, but that in the year this article was published, a subsequent record was set.

"In order to avoid going past the two-degree-Celsius temperature rise since the preindustrial era, the carbon budget would probably have to stay under eight hundred gigatonnes. A gigatonne is a billion tonnes—that's with nine zeros. And these are metric tons of course, heavier at almost 2,205 pounds than the American short ton at 2,000 pounds apiece.

"I suspect you've already done the back-of-the-envelope calculation, namely that releases on the order of forty gigatonnes a year would use up that lower carbon budget in approximately a decade.

"But who would want to do that? The answer is obvious. The fossil fuel corporations and countries in the world hold proven reserves of 2,900 gigatonnes, and yet most of them—whatever their rhetoric—seem prepared to extract and put into the atmosphere and ocean the carbon from every last lump of coal, every last drop of oil, and every last whiff of methane gas. Even worse, what are termed 'remaining ultimately recoverable resources,' or RURR, pencil out to nearly eleven thousand gigatonnes. This slide puts it all into perspective."

We didn't really need another acronym, Jake thought.

Excess Carbon in Fossil Fuel Reserves

- Proven reserves of 2,900 gigatonnes
- 2,900 divided by 420 ~7 times allowable carbon budget
- 420 Gt carbon budget
- 420 Gt divided by 42 Gt per yr. ~10 years
- RURR of 11,000 Gt
- 11,000 divided by 420 ~26 times allowable carbon budget

"This explains climate scientists' concern at the beginning of this decade, namely that to stay under a 1.5-degrees-Celsius temperature rise (equivalent to 2.7 degrees Fahrenheit), we needed to accomplish unprecedented changes in the next ten years." His voice rose on his next words.

"Any fossil fuel resources exceeding the budget must be permanently abandoned as stranded assets.

"Most people will never visit the Arctic. But everybody, whether or not they live on a coast, should worry about sea level rise—so this is one of our more persuasive arguments in favor of caring about climate change in general.

"Best data we have now is that relative to sea level in the year 2000, we can expect a one-to-four-foot further elevation of the sea by the year 2100. And emerging science on Antarctic ice sheet instability suggests that a rise exceeding eight feet is physically possible. Clearly, the nine to ten inches we have experienced since the preindustrial era is just a beginning. Before that, the geologic record shows an impressively stable sea level for the preceding two thousand years. But look at this daunting slide demonstrating what prehistoric sea level records should be warning us about."

Paleoclimatologic Sea Levels

- 1°C warming already represents long-term commitment to > 20 ft
- 2°C warming represents a 10,000-yr. commitment to ~80 ft
- RCP 8.5 (high scenario) emissions result 10,000-yr. commitment to 125 ft

---

"I know these numbers give us a sense of urgency, but I will now show you three lines of evidence that should provide a sense of hope, beginning with a trifecta of international climate agreements *that are actually working.*" Mikhail wrote on the board:

Montreal Protocol

Hope in messaging, Jake thought.

"In 1987, the Montreal Protocol went into effect, addressing not just ground-level but stratospheric ozone depletion. The United States saw Senate ratification and presidential signature during the Reagan administration. The refrigerants they banned were chlorofluorocarbons or CFCs. These leaked from cooling systems during production, transportation, and injection into appliances like refrigerators and freezers, during service life, and finally, through demolition or recycling of components. These chemicals are persistent and potent, because their heat-trapping effect is a thousand or more times greater than that of carbon dioxide. Chlorofluorocarbons were destructive not just to high-altitude ozone but also the stability of the climate.

"But a new refrigerant class of chemicals was developed, called hydrofluorocarbons or HFCs, and used in air conditioners, refrigerators, and heat pumps. In 2016, after seven years of discussion and diplomacy, negotiators in Rwanda's capital of Kigali reached a binding accord by more than 170 countries. Binding, because this agreement became the Kigali Amendment to the Montreal Protocol, and acquired the immediate force of international law.[51]

"The political compromise meant instituting three tracks of compliance. The richest countries had to stop production and use of hydrofluorocarbons in 2018. The majority of countries committed to halting HFC use by 2024, implying that most countries have crossed the divide. But a small group of the hottest countries insisted that they be able to defer compliance until 2028. These latter countries included India, Pakistan, Iran, Saudi Arabia, and Kuwait." Again, Mikhail erased and scribbled:

CORSIA

"The second salient, or noteworthy, agreement dealt with aircraft emissions. Article 2(2) of the Kyoto Protocol to the United Nations Framework Convention on Climate Change—quite the mouthful—charged the International Civil Aviation Organization or ICAO with responsibility for crafting aviation reductions—an organization conspicuously omitted from the Paris Agreement of December 2015.

"The ICAO is a UN special agency.[52] In September 2016, it adopted a standard for governing its carbon dioxide emissions. The agreement, with the typically turgid bureaucratic title of 'Carbon Offsetting and Reduction Scheme for International Aviation, or CORSIA,'[53] was coupled with airframe and engine design prescriptions and proscriptions." Mikhail shook his head once again.

"The following month, in October 2016, the first design certification standard for any industrial sector of any type was created. By 2020, the standard applied to new aircraft designs. And I'm happy to tell you that by 2023, this standard began applying to preexisting, or legacy, designs still in production.

"Aviation has climate impacts coupled with health risks. The reality of over one hundred thousand global flights daily speaks to the crux of the climate impact—that is, until a dastardly virus swept around the world and grounded millions of passengers in their tracks. Airplanes are contributing about five percent of heat-trapping emissions but are projected to double by midcentury. Part of the reason for the outsize effect of commercial flight is that the major release of jet engines by weight is water vapor—a product of combustion—high in the stratosphere, where the water molecules slowly morph into extraordinarily high-altitude and unusually persistent cirrus clouds exhibiting formidable heat-trapping effect."

Jake wrote down the data he knew would support his ongoing debates with Abbey about plane trips.

"Realize four out of five of the world's people have never been on a plane, and the majority of Americans do not fly in any given year. There really are frequent flyers, too many of them Americans and Chinese. This next slide enumerates the adverse effects of flight for the general population and the globe."

Health Effects of Commercial Aviation

"Unregulated emissions from planes ... above 3,000 ft responsible for most ... deaths."[54]
- lead author Steven Barrett, MIT aeronautical engineer
- plane crashes kill ~1,000 people annually
- particulate aviation emissions kill ~10,000 people each year

Aviation Decarbonization

- jets now 80% > fuel efficient than first jets 1960s
- algal, *Jatropha*, *Camelina* biofuels reduce life-cycle carbon up to 80%

"Now, for the ironic corollary: Climate change adversely affects safety in aviation.

"Somehow, the climate's effect on aviation seems to create a more level playing field, since the small minority of people who fly are helping to create havoc for over eight billion people—not to mention all the other creatures on Earth."

Adverse Effects of Climate on Aviation

- increased in-flight turbulence
- heat waves will cancel more takeoffs due to decreased lift in hot air
- sea level rise will flood many low-lying airports

"I should spell out that emissions under three thousand feet affect our health as well, but we regulate these emissions to some degree. The airline industry has experimented widely with the use of biofuel for aviation through a number of proof-of-concept flights, but electrification for aviation promises to be more effective and scalable.

"Next, Phoenix, Arizona, offers a good example of a city whose airports have had to cancel takeoffs but not landings when ground temperatures exceeded one hundred degrees Fahrenheit by too great a margin, owing to a fully fueled plane's greater weight at departure.

"And Miami, Florida, boasts a prime example of an airport that's dead certain to be affected by a rising ocean, only eight feet away from inundation." Jake tapped his fingers restlessly as Mikhail wrote:

Paris Accords

"The third general international agreement of note is the Paris Accords of 2015—after twenty years of convoluted negotiation. The agreement's two main impediments are the absence of enforcement mechanisms and the last Washington administration's attempted withdrawal. Obviously, the landslide electoral outcome of several years ago temporarily solved this problem. We are now moving forward with strengthened and enforceable treaties and greater international cooperation. That's right: cooperation, inclusion, and justice. A rejection of nihilism."

Jake made a fist with his left hand as he kept writing with the right. *Damn straight.*

"We have no choice if we're to 'have a fighting chance to head off the blistering heat, droughts and wildfires that are the hallmarks of a fast-warming planet,'[55] as Somini Sengupta cogently stated in the *New York Times* in May 2020."

"Worldwide, countries are stepping up in the nick of time to avert the crisis of *stranding* oil and gas assets, however painful this may be for the companies involved. Look at this sadly short listing of global initiatives."

Countries Confronting the Challenge

- Belize + New Zealand banned all offshore exploration
- Denmark stopped oil exploration onshore permanently
- France + Costa Rica halted onshore + offshore exploration + extraction
- Canada + US relinquished Arctic offshore oil + gas [but in 2023, US opened large parts of Alaska's North Slope]

"In short, continued exploitation of fossil fuels is no longer acceptable. We're finally ready for a paradigm shift in attitudes. We shall no longer believe that corporations are people until and unless we see one or three of them put into prison.

"We felt it important that we end this presentation on climate change on a positive note. After discussions with Jake and the other faculty members, and since this is the last presentation until after winter break, we also decided to allow ample time for questions at the end. Max—come on up here and let's open this up for debate."

Max stood, hoisted up his pants, and joined Mikhail at the front.

Rachel raised her hand first. "Dr. Baerbock, you're in the physics department. So does that make you more of an optimist or a pessimist about the whole climate story?"

Max paused, arching his eyebrows while considering the question. "From the physics of the atmosphere and ocean, we could, in theory, rapidly halt our release of heat-trapping gases, cease our clearance of virgin land, especially tropical and temperate rainforest—"

Mikhail threw in: "Which is about eleven percent of the carbon story . . . sorry, Max."

"No problem—yes—our focus should not be on greenhouse gases alone. Domestic animals on the hoof should become a much smaller part of all our diets. I am speaking here of cattle, sheep, goats, and so forth. In other words, from the science alone, we could still accomplish a rapid shift in direction."

"Of course a halt alone is insufficient," Mikhail added. Since we're speaking hypothetically: If we stopped releasing greenhouse gases tomorrow morning—since it's too late to do so today—the planet would go on heating up for three to five years, but then at a gradually slowing rate toward equilibrium over centuries.

"I will confide that I rarely talk about the serious changes accompanying us into the next century, as most people inordinately discount the future. That is, in public speaking I don't. I focus on this century only, while as a geologist, I concentrate on long spans from the past and on into the future. I'm committed to getting out of the ivory tower and engaging with the media and people in general. But I fixate first on the interval up to the midcentury, and second, on the latter half of this century."

Looking up toward Abbey, he said, "Dr. London, let me ask you a question. The world reached over eight billion people at the end of 2022. This begs a provocative question: With your medical background, do you feel we can halt and reverse population growth in the coming decade?"

Abbey pressed her fingers thoughtfully against her lips, then replied. "In the context of globally improving access to basic medical care, and given our reliable, long-term modes of contraception, I would say of course we could—over this century—get down to a more manageable population of seven billion or so. But longer term, I think a population of three billion would be a lot less likely to exceed the carrying capacity of the planet."

Mikhail nodded. "The world was at two and a half billion as recently as 1950. This begs a provocative question: Given the pressures of population growth, what are the ethics of having children in this day and age?"

Abbey replied just as seriously. "This country and China and dozens of others are at less than the replacement rate of fertility, so I have no question that having one or two children is acceptable and ethical. That said, I think we've got to remove most governmental pro-fertility incentives, simultaneously coupled with offering federal and state support for preschool and kindergarten through high school for all."

"By pro-fertility incentives," Max queried, "you mean tax breaks and such?"

"Exactly," she said, "deductions for the first two kids, but not any more after that."

A woman in the back raised her hand tentatively. "Dr. Ligachev, where are you on the optimism–pessimism spectrum?"

"As a geologist, my fundamental philosophy is that the Earth will go on orbiting the Sun for billions of years. Not forever, of course, but as much as four billion years before the Sun expands out and radically transforms the four rocky inner planets. I recognize that as a species we're altering the status of the troposphere and crust, but not in a way that cannot, ultimately, be reversed. I think you'll find most climate scientists exhibit an underlying optimism—because otherwise they would be too disheartened to carry out these scientific investigations, and would need to go into another line of work. You can count me at the optimistic end of the spectrum, in spite of my German and Russian roots."

Brian was next. "Do we place emphasis more on mobilization and mitigation of climate change, or retreat and adaptation?"

Mikhail wavered for only a moment. "It's too simplistic to say 'all of the above,' and simply stop there. Even if all trends were abruptly flatlined at the current level, we're still going to see increased storminess, heightened wildfires, more droughts and floods, sea level rise, ocean acidification and deoxygenation, algal blooms, extinctions, and coral bleaching. So we must put maximum push into mitigation to reverse all of these changes. That said, the ocean will rise, inexorably at this point, and we might be looking at eight to nine feet by the end of the century. Obviously, we're also going to carry out adaptation and substantial retreat from most of the coastal areas. Only some cities can be protected. If these projections play out, New Orleans and Miami will not make the cut, sadly."

"The question before us," Max added, "is can we be successful in all these endeavors to the point that the Anthropocene will look more like the Holocene epoch than the Eemian epoch? Ultimately, we need to bring the deniers and disbelievers on board with the rest of us—we're not leaving anyone behind. To Mikhail's list, I would add increased geopolitical unrest and the spread of tropical and emergent diseases, such as coronavirus, dengue, Ebola, Zika, and others.

"Interesting time to be a specialist in infectious disease," he added, smiling at Abbey, whom Jake knew was in complete agreement with Max.

"I have one last thought to leave you with," Mikhail said. "A quotation from Peter Raven, president emeritus of the famous Missouri Botanical Garden. 'It's up to us to decide what kind of world we want to leave to coming generations—a sustainable one, or a desolate one, in which the civilization we have built disintegrates rather than builds on past successes.'"

Addie spoke up ardently. "If I may, I think I'd add that one of our values must be that we protect the wild places on the Earth. One of my favorite lines is from John Muir: 'Climb the mountains and get their good tidings. Nature's peace will flow into you as sunshine flows into trees.'"

Jake looked around, stood up, and said, "Please now put your hands together to show your appreciation for our two fine speakers tonight." And the crowd indeed raised the roof, continuing their applause so long it sounded like the last night of a three-day rock festival.

TWENTY-SEVEN

Nosocomial Dreadnought

It's true most dilemmas are figureoutable,
but sadly the most intransigent are unfigureoutable.
—Anonymous

A bbey turned away in shock, trying to maintain her composure. She was suddenly trembling, her palms sweaty, her eyes closed for a few seconds to prevent tearing up, not trusting herself to speak. *Oh, fuck, fuck, not my baby. Please . . . not my baby.* She understood she had to pull herself together. She of all people knew the enormity of what had just happened, knew she had to put herself immediately into protective isolation and endure whatever other steps she had to take. But she couldn't stop her heart from pounding.

She took a deep, shaky breath. "Maggie," she called out as she turned, "I just stuck myself with a needle, damn it . . . I can't believe I was so stupid. Only subcutaneous, I don't think it went deeper. But it was so fast, I'm not sure." She heaved a sigh. "My left dorsal hand, I'm washing it with chlorhexidine now, really scrubbing. I need some help, a dressing."

Maggie was assisted by others as she struggled as fast as she could into PPE, then slid open the door to the ICU room and came in, her face pale above the mask. "Let me see, Dr. London . . . let me see what happened."

They both looked at the small wound between Abbey's index finger and thumb. Abbey had instinctively stripped off her left glove to express a few drops of blood from the wound and scrubbed the whole area to a delicate roseate color.

"Let me get a translucent dressing on this, for observation," Maggie said. "Our head nurse will notify infection control."

"Who . . . who's on tonight?"

"Duncan Harrison—he's in the hospital. I happen to know he's down on the second floor." She paused. "I want you to wait here until we decide what to do." Another pause. "Where to put you . . . you know we can't let you go home." Maggie looked through two layers of splash protection between their two pairs of eyes. "You have to be quarantined with blood and splash precautions until we see how this plays out. Wait here. I'll be right back."

Abbey had known what Maggie would say. Hospital protocol was very clear: Any person who has suffered blood exposure from a hemorrhagic fever patient must be quarantined until he or she is clear of signs of infection—well past the incubation period. *I helped develop the damn protocol.* She watched as three and then four of the staff gathered at the nursing station, with other personnel anxiously glancing her way. The patient she'd attempted to draw blood from was minimally responsive and on a ventilator. The family had given consent for a trial of a treatment investigational new drug, which had just recently been cleared through the hospital's Institutional Review Board. The protocol for the agent, CDG1529, entailed a series of blood tests prior to and following each infusion, as well as a host of other observations.

The nurses were understaffed, no phlebotomy tech was available, and the pharmacy had just delivered the drug in a small infusion bag with tubing covered in aluminum foil. Once the agent was mixed and ready to be administered, its photosensitivity and time-sensitivity meant it needed to enter a patient's bloodstream right away. Since Abbey was overseeing this study subject, she had decided to do a quick blood draw to facilitate getting the infusion started.

Duncan Harrison had arrived, and after first knocking on the door and smiling at Abbey, he was suiting up in his own protective gear. One of the nurses slid the door open and closed it behind him. He kept his voice low and calm—professional—even though he knew her well. "I understand you've had an incident, Abbey. I'm so sorry. Let's see if we can figure out what happened. Please start at the beginning."

Abbey had a small bandage over the puncture, and she was holding it up like a puppy with a sore paw. "I was using a butterfly needle, twenty-gauge. When I was done with the blood draw, I flipped it over and was about to snap the cover guard in place over the needle when I almost dropped the tubes of blood in my left hand. Without thinking, somehow I jerked across and pulled the butterfly tubing with me, and it ended up here," she said, pointing to the small wound.

Duncan examined her hand without touching it, then looked up. "How far along is the pregnancy?"

A quick inhalation, then she said, "Thirty-three weeks and three days."

"I bet that was your first thought," he said.

Abbey turned a shade paler but held back tears. *I can cry later*, she thought.

"I know you've had all the required vaccinations, including needle-stick-pertinent coverage against hepatitis B and tetanus. Still no vaccine for hepatitis C yet, after all these years. Is this patient positive for hep C, or is it just assumed, given his history of injection drug use?"

"He's seropositive of course, but has never received any treatment for the identified infection."

He sighed. "Do you have any identified drug allergies or idiosyncratic reaction history?"

Abbey shook her head.

"I'm going to ask that you accept admission for observation under protocol. I understand the nurses are readying an ICU bed down at the end of the row. This is, after all, the most logical place to house you, at least for the moment. I'll make all the arrangements for admission. Do you have a preference for an admitting doctor?"

"I'm receiving care from a midwife with privileges here. There are two obstetricians in her group—"

Duncan interrupted. "I'm going to suggest instead a perinatologist, Noah Bookbinder. English guy, you may know him."

"I don't know him . . . but I agree that a perinatologist would be appropriate. But only after notification of the obstetrician on call for my midwife's group."

"As soon as we get you moved over, we'll draw some blood. We need to know your baseline blood counts, liver and kidney function, blood type, and baseline test for the CCHF virus. Probably some other lab, but that'll be up to the perinatologist. Am I safe to assume you've been tested for hepatitis C and are negative?"

She nodded.

He looked right at her and said what she knew was necessary. "I suspect Dr. Bookbinder may want to do a blood type and cross-check, in case blood products become necessary down the line."

She studied her wound. "I know . . . I know that might become necessary, however unlikely. I also know we have no data on the safety during pregnancy of our experimental drug against hemorrhagic fever, so I would refuse to accept it.

"And I know my baby is too early to deliver on an emergency basis without an adequate indication." She looked at the back of her hand again. "The . . . National Library of Medicine had an article which recommended use of ribavirin in health care workers with a needlestick injury. But the FDA categorized it as pregnancy category X, so again I can't use it. My only hope is for hyperimmune globulin from patients who survived the disease . . . for that matter, we might want to consider hyperimmune globulin for hepatitis C . . .," she murmured and trailed off.

"I . . . you . . . Dr. Bookbinder can check with the CDC and get further recommendations."

She turned to look back at her patient. "The blood samples I drew are on the bedcover. Someone should transport them down to the lab and get the drug infusion started for this patient. My partner, Dan Phillips, will take over his supervision tomorrow. I'll let him know."

"I'll get right on that," Duncan said. "And I'll contact Dr. Bookbinder immediately." He paused. "We're all gonna be rooting for ya, Abbey."

Maggie returned as Duncan was on his way out. "Dr. London, we're ready for you now. We're going to have you keep your PPE on until we get you to the other room." She led her to an ICU room at the end of the corridor. Maggie helped Abbey strip off her protective gear, then soberly closed the surrounding curtain and handed her the hospital gown.

"Here's a bag for your valuables, except of course you'll want to keep your phone . . . you probably have some calls to make. The phlebotomist will wait to see what else your new perinatologist will want to order. Probably not an IV at this point, just a complete history and physical. Duncan will let you know as soon as he reaches Dr. Bookbinder. I've met him before; he's a good guy." She pondered for a moment. "I know it's late, but do you want me to order up a meal for you?"

Abbey tried to get a grip on herself, waiting a couple of seconds, unable to make a decision. Maggie said she'd bring her juice and possibly cottage cheese.

First, Abbey called her partner Dan and broke the news. Her voice was tight, but she didn't break down. He offered to come in right away, but she suggested he wait until tomorrow when he came in to look at their study patient. He'd heard that New York City had just received some hyperimmune serum and said he would immediately get on that, beg them if he had to, to get it sent out on an emergency basis. They worked out all the details, and the call ended.

It was ten o'clock, but Jake was used to her odd hours. She knew he would be home after the lecture, but not yet asleep. He answered on the second ring.

"Jake, honey, I have something to tell you. I want you to sit down and stay calm. This will probably not turn out to be anything serious, the odds are against that. I sustained a contaminated needlestick tonight . . . I'm so sorry." Her voice sounded shaky in her ears. "So . . . I have to stay here . . ."

"But, why, Abbey? Why can't you come home? What happened?"

Her throat was so tight she almost couldn't speak. "It was from one of my hemorrhagic fever patients, Jake." Her hands trembled as she gripped the phone. And then came her flood of tears.

Noah, in his thirties but balding prematurely, stopped a moment to rub his eyes before entering his admission orders and simultaneously calling Dan.

"Hello, this is Dan Phillips. What can I do for you?"

"Dan, this is Noah Bookbinder. I just finished examining Abbey. You can tell she's a strong woman, but in reality, she's scared to death. I talked to her partner Jake afterward and tried to calm him down. We're going to break protocol and have the nurses get him into gear so he can visit, as long as there's no direct physical contact. Just today, just one time, not any subsequent days.

"I'd certainly appreciate any advice about managing her care that you can give me," Noah said.

"I just got off the horn with some folks in Mount Sinai Hospital. I had to check around, but they have experience with this Crimean-Congo hemorrhagic fever virus—and some hyperimmune serum they managed to get from Israel last week. It took a bit of arm-twisting, but once I told them our situation, they said they'd get it on a flight and here within a couple of hours."

"Boy, I'd like to get that into her immediately when it comes in," Noah said. "Any other specific therapy we can offer her?"

"Given her exposure and the pregnancy, I'd recommend some hyperimmune globulin for hep C. I'm checking my contacts for that as well."

"IV or IM?" Noah asked as he rubbed his eyes again.

"IV for the CCHF, IM for the hep C. As far as monitoring for infection, I'd suggest a daily draw for the hemorrhagic agent, the one we're more worried about," Dan said, shaking his head even though he was on the phone.

"What does this disease look like?"

Dan deliberated for only a second. "This is an RNA virus usually transmitted by ticks. Incubation period with a direct contact like this ranges from five to nine days. Once symptoms start then we know the PCR should be reliably positive in the blood. What we don't know is whether the PCR and other tests turn positive before onset of symptoms, but I'd still suggest starting tomorrow morning as a baseline. Since it's an assay of RNA, there won't be any confusion based on all the antibodies you'll be giving her."

"Worst-case scenario, if it turns out she's infected, what does it look like?"

There was a slightly longer delay. "Acute onset of chills, fever, weakness, headache, myalgia and arthralgia, abdominal and lower back pain. As a perinatologist, you'll be particularly sensitive to that of course. Nausea and vomiting, photophobia, and reddened face, neck, and conjunctiva and pharyngeal mucosa are common.

"The fever is biphasic, with a sudden drop in the middle to subfebrile, then spiking again at the start of bleeding complications. Petechiae anywhere on the skin and in the throat, with large areas of ecchymoses or other bleeding scattered over the whole body. Bleeding from the nose, blood in vomit, urine, and stool." Dan paused. "Bleeding from the genital tract even if not pregnant."

"Laboratory findings, Dan?"

"Pancytopenia really, especially a drop in white cells, but also red cells and platelets—hence the bleeding risk. Elevated sed rate as just a nonspecific marker of inflammation."

"I've not had a chance to look at data on pregnancy. Whadaya got for me?" Noah said.

"Reports in the literature on forty-two cases, mainly out of Turkey, Iran, and Russia. I'm sorry to report results of thirty-four percent maternal mortality rate and a fetal mortality rate of fifty-nine percent." Dan was quiet for a long moment.

"Christ," Noah said. "Well, this is a healthy young woman, Dan. If she is infected, I think we can do better than that. And the antibodies we give her may lower that risk appreciably. Let's get right on this. I'll finish my orders and get my admit note dictated."

"I'll have a chance to meet you in the morning. And I need to round on that other investigational drug patient as well. I don't want to lose my partner, Noah. Let's give this a full-court press."

"We're moving as fast as we can. My orders will specify I get notified as soon as these deliveries come in, and I'm having the nurses start an IV now so we'll be ready. Presumably the first PCR tomorrow will be negative, but my hope is that all the subsequent ones will be as well."

"Fingers crossed. See you tomorrow."

TWENTY-EIGHT

Calendar Days

Every bad situation is a blues song waiting to happen.
—Amy Winehouse

*Perhaps all the dragons of our lives are princesses who are only waiting
to see us once beautiful and brave.*
—Rainer Maria Rilke, Letters to a Young Poet

Brian and Rachel sat on a couch in their carriage house apartment in late afternoon with their communal laptop, dialing up a video call with Abbey. Wintry blasts rattled their living room window, but they were snug as bugs with a blanket across their laps, each nursing a cup of hot chocolate. They'd been denied access to their friend in special isolation at the hospital, but Jake had been getting their food deliveries daily, with enough for him to share.

The poor guy had lines on his face they'd never seen before, juggling his work schedule as best as he could to visit Abbey twice a day through the glass. Jake told them he hadn't been sleeping well, not even six days out from her exposure. Her blood tests had stayed normal, including the PCR, along with the blood counts and liver function tests. Jake had rapidly become conversant with these numbers, obsessively communicating them to friends and faculty at the school. Everyone tried to keep his spirits up, but his tension was palpable to all.

Abbey's face suddenly popped up on their screen. Rachel held both hands behind her ears to encourage her friend to turn on her audio. "Abbey! How're you feeling?" Brian said. Everyone knew by now the early symptoms of the lousy virus. While they were trying to sound casual, they were clenching their hands at the same time, waiting for news that she had no chills, fever, headache, or other warning signs.

Abbey knew in her heart of hearts she had the task of trying to reassure her friends that so far, she was okay.

"I'm good, guys, how are you?"

Rachel and Brian visibly relaxed a bit hearing her sound so steady.

"All my tests today are fine, though it takes time to process the last PCR result. But I had a repeat draw this morning." She held up her arms to show her multiple band aids, and they could see she had no IV. "Tell me, how do you think Jake's doin'?"

"I won't sugarcoat it, Abbey," Brian piped up. "He looks like shit. Let me take that back, he just looks exhausted. And I have to say you look pretty fatigued as well."

"I'm actually sleeping a bit better. It's noisy in here with everybody's equipment giving off beeps, but I'm used to it. I woke up to the adjacent room having a code blue last night. My first response was to react, but instead I rolled over and finally fell back asleep."

"How did the code blue go?" Rachel said.

"Better not to ask," she said calmly.

Rachel mumbled, "Whoops."

"It wasn't your experimental patient, was it?" Brian asked.

"No, I can't really say anything because of confidentiality issues, but I'll just say he is still with us." All three of them seemed to let out a collective breath.

"Abbey, Jake was saying something about hepatitis C? I hadn't heard about that before. Explain, please," Rachel said.

She nodded authoritatively, just once. "Injection drug users usually end up sharing their needles and drugs, and almost all of them end up getting infected. Its incubation period is longer than hepatitis A and shorter than hepatitis B, if that helps." She could see Brian scrunch his eyebrows, and Rachel shook her head uncertainly.

"At any rate, hepatitis C would not be good in pregnancy, but much less serious in a way. There is less urgency about identifying this, and most of the tests are antibody-based. The reason that's important is they gave me a lot of gamma globulin in the first twelve hours after I got here, first for the hemorrhagic virus, second for the hepatitis virus. Since each is pooled from hundreds or even a thousand patients, it's certain I received some antibody for hepatitis C—"

Rachel sat up straighter. "Aren't you worried about being exposed to blood from so many people?" Abbey was already shaking her head.

"No, no, these preparations are sterilized three different ways, so no concern at all. Just a collection of antibodies."

"So, how do you get around this for testing?" Brian said.

"The Crimean-Congo virus is extraordinarily rare, and hepatitis C virus is common. Probably at least 2.4 million people live with hep C in the US. So in both cases we're testing directly for RNA of the virus, which obviously would not be in these special antibody concoctions."

Rachel and Brian appeared to mull this over.

"As you already know, we're testing daily with PCR for the hemorrhagic virus, and we have been talking with everybody about how long is sufficient to feel confident I've escaped a bullet. The NIH, the CDC, and several other med centers have weighed in, not to mention some folks at Oxford. I'm having a number of other tests and lots of saved blood for future potential assays. There aren't a lot of case reports of nosocomial infection—"

Rachel screwed up her face. "What's that?"

"'Nosocomial' just means acquired in a health care setting. An occupational risk for all health care workers, but occasionally other patients and visitors. Anyway, with little information on this peculiar circumstance, I have happily turned myself into a guinea pig." *Gives me a sense of agency, I suppose.*

"We all feel more hopeful every day that passes. The consensus is that seven days of negative test results should suggest I'm okay. This means that on my eighth day in the hospital—when we get the seventh-day result back—I may be discharged. Both Dan Phillips, my work partner and my perinatologist agree with this plan."

"Then you're home free?" Rachel asked.

"No, not yet. But the odds of infection will have fallen low enough that I can reasonably stay at home. Before discharge I'll get my first PCR for hepatitis C, as that usually turns positive one to two weeks following exposure. Or needlestick in my case. So that'll get repeated for both viruses every other day for ten days with a home visit by a visiting nurse serving as my phlebotomist."

"Can we visit you then?"

"No, you can't," she said immediately, and she could see their regret, though perhaps mixed with a bit of relief. "Jake has canceled all visitors to the Inn for the next twelve days. He'll sleep in an upstairs bedroom for the duration, and I'll mainly stay in the bedroom downstairs. After a couple of days, I might be able to gravitate to the kitchen. If you want, I could bundle up and come out to the porch and wave at you at least."

Rachel started to cry, and Brian, surprised, hugged her tightly. He couldn't speak either, so Abbey did.

"Rachel, it's okay. I know this has been tough on all of us. I feel like I've cried enough tears over the last week to last me a lifetime. But we've got to be resilient; this isn't over yet.

"Noah, my perinatologist, has the nurses run a monitoring strip on the baby each morning before he comes in, and he's also done one bedside sono, but there's nothing else to do but follow the literature. He has a great smile, by the way. He knows as well as I do that I'm not out of the woods yet, but it's looking better every day. My midwife came in for a visit once and was trying to be exceptionally reassuring. She listened to the baby too. It was so good to see her."

Rachel wiped her eyes. In a hoarse voice, she asked, "Did the sono confirm the baby's sex?"

"Yes, it did. But it's kind of hard for a tiger to change its stripes, right?"

"I won't bother to ask the obvious question," Rachel said, sounding resigned.

"And I will neither confirm nor deny anything," Abbey said with a smile.

Brian was more stubborn. "Come on Abbey, you can trust us. Is it a boy or a girl? We won't tell Jake."

"Yes it is," she said. Nothing more, nothing less, still smiling: a boy or a girl.

TWENTY-NINE

Pandora's Box

We wait until Pandora's box is opened before we say, "Wow, maybe we should understand what's in that box." This is the story of humans on every problem.
—*Peter Singer*

I t was two in the morning and all the critical readouts were nominal at Peach Bottom Atomic Power Station in Delta, Pennsylvania, sixty-eight miles northeast of Washington, DC. The plant used cooling water from the Susquehanna River at a point three miles north of Maryland. The nuclear technicians were superbly trained and attentive. Most of them had only been on shift since eleven that evening, but nearly all were night owls and adapted to the schedule.

It's hard to recognize something when it's not there. The ambient sound may have been quieter inside and different in pitch outside—but only to an observant ear.

The reactor status held stable at ninety-eight percent full power. PJM, the largest regional system grid operator in the country—with coverage including Pennsylvania, New Jersey, and Maryland—had made no requests for any changes. Aside from routine traffic, no requests had been submitted for information. No transmission anomalies appeared in the adjacent grid area.

A nuclear power plant contains hundreds of valves, some more critical than others, and many external to the containment building. Now several of the auxiliary valves supplying the heat exchanger had slowly started to open. But the flow and pressure measurements stayed reassuringly—misleadingly—normal. At the same time, the initial dripping and vibration escalated to high-pressure, high-temperature steam blasting out in a high, piercing whistle, which would build up louder and louder until it sounded like the roar of an approaching train.

Hundreds of switches and circuits were arrayed intricately in displays at all of the control stations and at other locations throughout the large structure. Unknown to the operators, a critical subset of these controls were uselessly fixed and unalterable.

The security guards on the perimeter were looking outward, inspecting doors, scanning the perimeter, but not attentive to the plume of steam from the cooling towers, which gradually attenuated. But as the whistling crescendoed and the leaks from the valves grew alarming, they tried repeatedly and unsuccessfully to call into the control room. One was deputized to run, not walk, and alert the control room supervisor.

An automatic computerized transmission from PJM indicated a problem with power drop, but the audible alarm inexplicably failed to sound. Seconds later, a direct phone call came in from the RSO. "This is Jim Stanley. Get me Herb right away."

Herb Stinson was an experienced hand, a little thick in the gut, flinty-eyed when angry, and balding on top, with no tremor in his hand or voice. He picked up the receiver at his central station. "Jim, what's up?"

"What kind of output do you have down there? We're getting you at ninety-two percent and falling. What does your instrumentation show?"

Herb's voice got a little gruff, but he was immediately alert. "I can't confirm that. I'm looking at ninety-eight percent, everything nominal. Why the hell this discrepancy?"

"Herb, we also have a transmission anomaly on the line serving the area on the eastern side of the capital. I'm going to call up some other assets right now. You'd better start troubleshooting at your end. Get back to me in no more than two minutes. If you need to power down units two and three, we need to coordinate this right away." A sudden click, then multiple automatic communications began streaming in on various computers.

Herb stood. "Listen up, everybody. We've got a situation on our hands. Start double-checking the status of all your readouts; particularly focus on online transmission and the temperatures and neutron flux in both reactors. You're gonna have to be skeptical; there may be cyber interference. Brent, call in some of the other senior staff and let's get this sorted out. Right now."

Backup monitoring systems were turned on, but cut out repeatedly in only seconds. Overhead monitors showed fluctuating pressure in units two and three, but temperatures read out as normal—which didn't make sense. Warning lights on other boards turned on, but audible alarms were silent until suddenly, multiple klaxons sounded off at once—a bedlam of sound.

Valves were opening in the heat exchange units that had not been opened since the last turnaround and refueling: eighteen months ago for unit two, ten months ago for unit three. Superheated water spilled out of the heat exchangers, first in a spray, and then in more serious amounts.

"Boss," Brent called out. "I can't reach Reynolds, Chandler, or McEwen. Nobody's pickin' up."

Jim called back in three minutes. "Herb, I said two minutes, not three. Things are clearly headed sideways. I'm hearing a lot of alarms going off there, but we don't have similar readouts here. And all three of your high-voltage lines out of the park are showing anomalies. We're going to shut down these lines, which if we are unlucky will brown out or black out an area, including parts of the capital. Time to shut down your two reactors until we get this figured out. Make sure your diesels are ready. The last thing we need is a station blackout. Jump on this. And this time call me back in two minutes. Hurry!"

Herb got on the loudspeaker to communicate over the alarms. "Full scram both reactors. I repeat: Full scram both reactors. Do it." His voice grew sharp on the last command. "Shut 'em down—now!"

Cadmium alloy control rods were rapidly advanced in unison in both reactors. One rod in unit three got stuck only halfway in. Handheld instruments were plugged in to independently confirm the scram status.

Fissioning was almost completely shut down in under a minute. But with the crippled heat exchange units, the cores were not cooling. They were getting hotter from all the fission products, already ramping up past normal operating temperature.

The first automatic emergency core cooling system was triggered but failed to activate.

"The first-order ECCS is not working!" Brent yelled. "We gotta go to the secondary backup system. We still have power from the backup diesels, but three of them on reactor two failed on startup."

Herb yelled, "We've got some core uncovery of the fuel rods starting, people. Let's fix this before we lose the geometry of the fuel rods." *If that happens*, he thought, *we'll lose control of the reactor and all hell will break loose. If the fuel rods start to melt, then we'll see hydrogen building up . . . a goddam explosion.*

Herb strode to Brent's station. "The book says even one diesel generator can keep a single core cool, as long as it has water to pump through the pressure vessel. Even if the second- or third-level ECCS functions, they can't last forever. Designate eight or ten people to troubleshoot the diesel generators and figure out how to goddam jury-rig them, or at least switch one of them over to reactor two. Hurry before core uncovery progresses. We are not gonna have another Fukushima here—not on my watch. These people have been drilled on this scenario. Make sure they run like the devil's behind 'em."

Dozens of men and women scrambled and spread over both reactors like a professional basketball team moving down the court. The opened valves on the heat exchange units were closed with separate controls and manual confirmation. Two men and one woman suffer steam burns, all three of them screaming and cursing. Herb knew they were going to need evacuation by ambulance.

The pressure began to stabilize and then dropped again, suddenly and without explanation.

Herb heard Brent yell in his direction. "We're losing control of reactor two. Core uncovery's getting worse, temperature's up to 1,180 degrees Fahrenheit. Dammit, looks like reactor three is not far behind."

"We've got a civil emergency on our hands now. I told Jim Stanley to notify the governor and get him briefed. If we can't stop this now, we'll need to evacuate. Iodine dosing, especially for kids and pregnant women. Releases of radioiodine, xenon, krypton, cesium-137 on our sensors is just starting, but so far catching only small quantities in filters. Nothing released off-site."

Watching his staff work feverishly, Herb was rewarded when a single backup auxiliary electrical line to outside generators patched in an hour later, allowing staff to commandeer outside support for the ECCS system. But they would only reverse partial core uncovery on both reactors hours later.

PJM, the regional grid operator, realized early on in this accident they could only assist Peach Bottom in the immediate emergency, but were going to lose their grid participation for many months. Fortunately, the grid operator had multiple resources and enough high-voltage, long-distance transmission to stabilize output in the grid. Wind farms from as far away as Wyoming would be part of the mix of generators for the first several days until the picture settled down.

Herb's shirt was sweaty in the pits. He rubbed his unshaven face at about noon the next day. He leaned on a desk and told Brent, "FUBAR all the way, Brent. We got lucky at the end. But the record will show not that we almost lost Detroit, but rather we almost contaminated Washington, DC. Christ, we were lucky. But we're gonna be spending a lot of time with NRC staff and no doubt congressional committees as well. Let's start pulling all the documentation together."

"Two reactors at the Peach Bottom nuclear plant abruptly shut down in Pennsylvania tonight," read the press release to the morning newspapers. "The causes are still under investigation. The localized blackout is slowly being reduced as lines are reenergized, and federal governmental functions were not seriously impacted. It is important to understand that the public was never in any broader danger. No radiation releases occurred in this incident, but both reactors will be out of service for an indeterminate period. The governor will hold a press conference at noon to provide further information."

THIRTY

Rapprochement

*You survived; I survived. We're together again. I once begged the gods to let
me see you—if only for a moment. To see you and know you'd made it.
Just once; that was all I ever hoped for.*
—*Sarah J. Maas,* Queen of Shadows

It had snowed again last night, well past the holidays, weather that had
now continued almost to the end of January. The Inn seemed so de-
serted without guests. Jake came down the stairs as quietly as he could,
counting each step in the dark and, from long habit, stepping over the two
creaky steps. He was in stocking feet and an old robe. He could hear Abbey softly snoring in their bedroom downstairs. *She doesn't think she actually
snores sometimes, but I know better.*

He turned on a light, finally, the one in the kitchen. He got the coffee
going, making only half a pot for him alone. With nothing particularly
planned, he just scrambled some eggs, adding a bit of hot sauce and a
generous amount of pepper leavened with a small quantum of salt. *Aha,*
he thought, discovering a loaf of rye bread and slicing off a few pieces.

Ironically, it wasn't the bubbling coffee maker or the smell of the eggs,
but the *sprang* of the toaster that finally woke Abbey up. He got out some
plates and set up at the small table in the kitchen. He heard the toilet flush
and knew she'd be out soon.

In fact, it was only a minute later that she shuffled out in her slippers
and robe, red hair tousled. She was rubbing an eye with one hand and
holding the other on her lower back, her abdomen shockingly rotund.

"Morning," he said, and she nodded. "Get some sleep?"

"Yes, I did, though the kid woke me up a couple of times. I never used to sleep on my left side, but it's quite comfortable now."

I know better than to make any more cracks about whales. "You are so gorgeous when you're pregnant," he said, smiling tenderly, speaking softly.

"You keep saying that, but don't stop even if I don't really believe it."

"So tell me what the Las Vegas odds are again."

"Can we sit down first and get some of those eggs and toast?"

"Have a seat. What kind of jelly do you want?"

"Blackberry, if we have any left. Butter's on the table."

Jake brought her a plate with heaps of scrambled eggs. Then he found himself kneeling on the floor next to her unexpectedly, and put a hand on her belly, waiting to feel a kick.

"There. Did you feel that, Jacob?"

Yes, I most certainly did.

"I can tell the baby is head-down, which is good," Abbey said. "Grab yourself a plate and I'll crunch the numbers again," she told him. "Like any betting enterprise, every player is working with incomplete information. Just like a hand of poker—"

"Texas Hold'em, or seven-card stud, or . . .," he interjected as he was standing, stopping when he saw her look of patient weariness. "Okay, forget I said that. So . . . incomplete information." He buttered up some toast and spread jam over it.

"I can still see the college boy in you," she said, shaking her head. "Yes—incomplete information. There aren't enough data in the literature, anywhere in the world. We don't know the risk of a needlestick. We don't know the degree of risk attenuation associated with a needle going through a glove, though with HIV, it's about fifty percent. We don't know if hyperimmune globulin is effective right after exposure—the way I received it in the hospital. And because of the antibodies I received, we don't know if my PCR test might have missed the presence of occult infection. We don't know what the risk to a pregnancy is."

But from her clouded look, he knew she was withholding some information about what she kept saying were indeterminate gestational risks.

"We don't know if the baby can catch the virus or is at risk solely because the mother becomes ill. After all, not many infectious diseases can cross the placenta, unusual things like toxoplasmosis or syphilis or Zika or . . . maybe I should stop there. We don't know if the antibody treatment works by preventing the virus from entering any cells at all or just targets those infected cells for protective destruction, what we call 'apoptosis,'" she said.

"What's that again?"

"When a cell is triggered to kill itself to protect the whole organism." She took another bite of eggs, mumbling that they were really good.

He smiled and slowed down. "The biggest question is, when will we know for certain? For really certain, Abbey?"

She shook her head. "Jacob, I wish I could offer you certainty, but I can't. What I do know is that I spent eight days in the hospital and now eleven days at home. The visiting nurse has been drawing blood every other day since I left the hospital, and every single test has been negative so far. My liver function and cell counts are all normal for pregnancy. There's no blood draw today."

"I still worry about your developing anemia with all these blood tests."

"But I'm not anemic, as I just said—we know that from the tests. And we know that a woman's blood volume increases by half by the end of pregnancy. So . . . stop worrying about me."

He just stopped chewing, tilted his head, and looked at her as if to say, "Yeah, right."

"As you know," she said, "you're going to get some blood tests too. And they'll collect cord blood from the placenta for more than just the usual assays. The scientist in me knows the case report of my needlestick is going to be a big deal. In fact, multiple publications will emerge based on this experience, with authors including me, my partner, the perinatologist, and people from the CDC, NIH, and Oxford. Not to mention the National Emerging Infectious Diseases Laboratories in Boston and the US Army Medical Research Institute of Infectious Diseases at Fort Detrick in Maryland.

"Did I leave one out? Oh, yes, the CDC's Building 18 in Atlanta. So, yeah, these will be multiauthor papers about many aspects of my care. This single report is unlikely to change medical practices, except that paradoxically it may become the standard of care for nosocomial needlestick injuries with these and other similar high-risk patients."

Several moments passed between them in silent contemplation.

"I didn't want to make my name in medicine with all of this, Jake, I really didn't." Her voice was getting tight. "I didn't want any of this . . ." Suddenly, she was bawling, tears rolling down her cheeks and dripping to her belly. He got back on his knees again to hug her, and they held each other for the longest time.

When she finally quieted, she looked out the window and stared. "There is a ray of sunshine out there at the moment."

Jake turned to see sunlight streaming through a small break in the snow clouds.

"And there's a ray of sunshine in here too."

"Whadaya mean, honey?" he said softly.

She'd sounded almost shy, but then she turned and looked at him and whispered, "I decided last night, after you'd gone to bed—and after I got the PCR report back from yesterday morning—that today I would finally let you kiss me."

So, he did. He kissed her on the back of her hand, then the palm, then lifted the sleeve to kiss her elbow, then pulled her robe down to kiss her shoulder and leave a trail of kisses up the side of her neck. When her robe fell further, he couldn't resist kissing the soft upper curve of her breast, then he nuzzled her nipple for only a moment before she pulled him up for an open-mouthed kiss, as if they had reached an oasis after a long, treacherous journey.

"Can I take you to bed now?" he whispered.

"Yes, but only to cuddle, Jake, not to let you inside me."

He thought for a moment. "But if your mouth is finally safe for me, I may have an idea for what could happen."

She looked at him long and hard, as if trying not to encourage him with a smile. "There's no literature on that, young man, but it's obvious you're quite eager." He could sense she was continuing to calculate the odds.

"Why don't we head back to the bedroom and I'll see what I can do, and actually what you can do for me too." Abbey said.

"Ask not what your country can do for you, a—"

She put her hand over his mouth and took him down the hallway to paradise.

THIRTY-ONE

Cryptological Coda

A lone wolf doesn't tread paths its ilk leaves; it makes its own footprints in the snow. Most of its kind lives in packs, but it is an army in itself.
—*Savas Mounjid*, The Broken Lift

A month had passed with the sense of a false spring thaw when the calendar page wasn't supposed to turn to the next season for almost four weeks. Icicles ran with single slow drips of purest water, catching sparkles of sunlight during an afternoon breeze a few degrees above freezing. The Dickensian slush in the streets was brown and gritty, with heaps of snow from the plows that looked like they would last until the real spring finally arrived.

Everywhere, people were wearing boots, leaving them grouped, still wet and dripping, inside doorways in schools and places of worship. The Inn had the boot room just for that purpose, nicely finished off in a utilitarian dark wood.

Dragonfly Inn was back in operation. The people who had been canceled had only been told that, for reasons out of their control, the proprietors needed to close for a couple of weeks. Many, though not all, had simply rescheduled for later dates.

Even Abbey's family had not been fully apprised of the situation, and instead Abbey assured them they need not come home just yet. She'd find time enough later to tell them the whole story, perhaps after the birth of their child.

But their previous guest, Emmanuelle, had called out of the blue this morning, asking if she could see them both at the Inn in the afternoon.

The hour had arrived and she showed up right on time, a woman of her word.

This should be interesting, Abbey thought, and answered the door. "Hi, Emmanuelle. Good to see you. I met you and Ben back—"

"It was September, Abbey, and as I recall, Jake said you had to immediately head to the hospital for some reason. We'd talked just briefly about our daughter, Margaux, and your pregnancy. Jake had a lot more time to talk with Ben and me in the morning and served us a memorable breakfast."

"Well, come on in and hang up your coat, here's the boot room," she said, pointing to the doorway.

Emmanuelle took off her anorak and footgear and came into the parlor.

Jake pulled himself away from his computer and came to greet her. "Howdy, Emmanuelle. How's it going? How are Ben and your daughter . . . daughter . . ."

"Margaux," she said. "She and Ben are just fine, though she is running all over the house now, naturally athletic and babbling to beat the band."

"Would you like something to drink? We have some apple cider already warmed up."

"That sounds great, but you don't have to go to any trouble. I'm not a paying guest this time."

Jake shook his head in mock rebuke and went to get her some cider while Abbey and Emmanuelle found seats in the parlor. When he brought their guest a cup and saucer, she took a sip and picked up her purse, removing two thin documents.

"Let me start with some background. But before that, I'd like you both to sign a nondisclosure agreement."

There was abrupt hesitation as Jake and Abbey looked askance at each other.

"What in the world for?" Abbey asked.

"Two reasons. But I cannot tell you more unless you are both willing to sign these documents. You should be aware of serious penalties for breaking an NDA with a government agency, including the potential for jail time. I talked to several of my superiors about even approaching you, but I can't say more unless you agree." Emmanuelle knew inquisitiveness alone drove most people into cooperating. She also knew from the prior visit, that she'd established a good rapport with the couple, a trusting relationship on which so much depended in her line of work.

"Okay," Abbey said. As a physician, perhaps she was used to making quick decisions every day.

Emmanuelle looked at Jake who was still crinkling his forehead before offering a decisive nod. She was relieved but not surprised and handed each of them a pen to sign the two copies, one to keep for their own records. "I have a lot of things to tell you, including my formal suggestion you keep these NDAs in a locked file cabinet."

"We have one in the back office where I work."

"But you don't always keep it locked, do you?" she said primly, leaving them both speechless.

"You've known for years we live in a surveillance society, so we need to get past this point right away. When we approach individuals—even those with whom we would like to develop an informal relationship—we must exercise utmost caution. You've had a number of visitors in the last several weeks, including two couples last night, and both rooms emptied out this morning. But during that interval—and unbeknownst to you—one of your guests accomplished a thorough electronic search of this house. You probably won't be able to guess which person it was, and I'm not going to say, because you have no need to know."

Emmanuelle knew they were trying to hide it, but the shock on their faces was obvious.

"Should I go on?" She knew this was a hurdle for them to get over, but curiosity is a great driver.

Jake and Abbey nodded.

"You don't even have a burglary alarm on this house. You must think this is a pretty safe neighborhood. I also know you have no guests tonight. I checked before coming over." She paused, looking at them intently, first one, then the other.

Jake pursed his lips and narrowed his eyes. "Did you have the National Security Agency check to be sure we were home alone this afternoon?" he challenged.

"Don't be foolish. I just confirmed before I drove over in this slush that your website listed all three rooms as available tonight. Don't make this more complicated than it has to be. Simple solutions often suffice. Got it?"

He nodded, looking embarrassed.

"Next, I need to tell you we've already gathered a lot of information about you two. Abbey, in particular: Your father Colby, long known to us, vouches for you without reservation. If we do agree to establish a special relationship with you, when he and Anne finally retire to the Inn, they may continue that arrangement with us.

"Jake, I will ask what you know about Mr. London and his actual position." She observed him with interest.

"I know he works as an intelligence agent, probably a COS, chief of station."

"How'd you figure that out, may I ask?"

"A guest one time was asking how the two of us, young as we are, could be running an inn. So I explained that Abbey's parents were off in the United Arab Emirates, and that he worked as a cultural attaché. The guest surprised me by saying that job title was often used as a cover for a CIA operative. The guest was an attorney in Washington, I figured . . . pretty well connected."

"I'd like to get his name from you later, unless you remember it now. How did you confirm your suspicion about Colby?"

"It wasn't until months later when I asked Abbey about it. All she could say was that she couldn't say anything and to please not ask her anything more." Jake gazed fondly at his partner. "We're in love, Emmanuelle, you know that, and she didn't wanna lie to me."

Emmanuelle lifted her chin and pursed her lips. "I should make it clear we're not recruiting you for the CIA or any other intelligence agency. Rather, we just want a secure place that out-of-town guests, perhaps, might need for a night or two. In some sense, you have a great cover here: a long-operating inn with a splendid reputation, seemingly above suspicion, especially given its location outside of DC proper. We'd much prefer to keep Abbey's father in his current assignment. He's a veteran operative and has been for years. Abbey, both your parents have gotten fluent in Arabic—no surprise there."

"Do you know my parents?" Abbey said.

"Only by reputation and one phone call with each of them," she replied. "That's the other task I wanted to talk to you about, Jake—to identify people in the engineering department or elsewhere at the university who might be interested in working for the government. You wouldn't recruit them. Instead, just pass their names along to us."

"Who's 'us'?" Abbey asked.

"It doesn't matter."

"Are you going to be our handler?" Jake said.

She suppressed a sigh. "Think of me as nothing more than a glorified contact you have in the government."

"Would we need any special training?" Abbey asked, perhaps remembering her parents.

"If you mean training in methods and materials—no. After we give you some time to mull this over, you'll need to schedule an appointment downtown to fill out some paperwork—a contract of sorts—though with neither pay nor benefits.

"Do you have any other questions for me now, the kind that I might want to or be able to answer? I sense a lot of questions swirling around behind your eyes, both of you."

Neither spoke.

"Now, Jake, I have a whole other issue to bring up, one where I had to be your champion in discussing this with my superiors. Abbey may stay here, though this concerns only you. Would you like to hear about it?"

The mood shifted. Emmanuelle thought this would be irresistible for him.

"Absolutely," he said.

"First of all, Abbey, we have no further expectations for you beyond what we've already discussed. We based this decision in part on your working milieu—not on the basis of recruitment, but on the importance of medical confidentiality. We couldn't imagine a scenario in which you could reveal information without compromising that core principle."

Abbey moved to speak, hesitated, then said, "I fully understand and agree."

Emmanuelle turned to face Jake directly. "Ordinarily, we never give feedback to sources of information, but after discussion, we decided to make a special exception for you." She took a slow gulp of her now-cool cider, then set the mug back down.

"No doubt you saw the newspaper reports of an event at the Peach Bottom plant. While your city of Clarendon was not involved in the electrical outage, those two reactors remain offline, and we're not even sure how to root out and defang the malware that crippled them. It's going to be a complicated process, because we've never seen an attack like this before. Incidentally, both these reactors suffered a partial uncovery and some limited meltdown, though nowhere near as extensive as what happened at Three Mile Island.

"And, Jake, we were impressed with the information you passed on to me some time ago. You had a gut feeling about someone acting in a shifty manner and had the gumption to contact me afterward. That's a pretty rare combina—"

Jake's eyes opened widely; the hairs on the back of his neck stood up. He gripped the handles of his chair. "Are you . . . are y-you saying that guy I saw actually had something to do with this?"

She nodded slowly. "His name is Greg Stenger, though that hasn't reached the newspapers yet. He was a sonar tech on a boomer—a nuclear missile sub—with a high security clearance, no demerits, discharged honorably a decade ago. Never anywhere near the power plant of the sub but conversant from contact with the crew. After mustering out, he completed an undergraduate degree at Columbia, focusing on mathematics and computer software. He was accepted at Princeton for a master's program in the same fields, but inexplicably dropped out after eighteen months and has had no further contact.

"He was an only child, a bit of a loner. Remarkably pale because he avoided sunlight, wore sunglasses any time he wasn't in the house that he was renting about a mile from the university. He has no visible tattoos unless he is in a sleeveless T-shirt, in which case he has a Navy anchor on his right shoulder.

"Mr. Stenger's profile fits that of a lone-wolf style of terrorist. But no political affiliations. No online contact with rogue political idealogues. When information reaches the newspapers, pundits will say he's a white domestic terrorist. But my personal opinion is that he doesn't quite fit the usual picture of a politically or racially motivated anarchist—like the members of the Proud Boys, for instance.

"So far he's not saying much at all. Certainly not anything about how he pulled this off. When the station blackout started at Peach Bottom, the short list of entities of concern for this geographic area were not what first came to mind. Instead, we looked at state players like Russia, China, North Korea, Iran, Saudi Arabia, but got no traction there. When the department came up short, we were quick to cast about for other ideas with great urgency.

"Mr. Stenger had tried to stay off the radar. But with a US Navy record, college attendance, and recorded video of him attending the lecture on nuclear power at George Washington University—not to mention following you out during the break in the lecture—we identified him within hours.

"With a court-ordered search warrant, FBI agents accomplished a forced entry the same night. He didn't put up much physical resistance, but the agents had to shoot his dog. Since he was on the third floor, he had just enough time to pull out his hard drives and get them into a crucible prepared for a thermite reaction with a short magnesium fuse."

"That burns like a son-of-a-gun," Jake said. "I saw that in a high school chemistry class."

"The agents couldn't believe he didn't burn the place down. He was visibly agitated, especially after they put the dog down. But he was quickly subdued. The fire department arrived after the blaze was already under control because the agents limited the damage with fire extinguishers."

"How did he ever pull this off?" Jake asked. "Though I guess I should preface this with 'allegedly' pulled this off."

"The department basically thinks he's a wizard. Not necessarily a genius, but extraordinarily bright. The first stage was phones," she said, shaking her head. "Have you heard of a system called 'Phantom'?"

Abbey spoke up. "I may have read a newspaper story about that a year ago," she said, rubbing her belly and taking a slow, deep breath.

"He managed to insinuate this malware into the phones of a number of employees working at the plant. Presumably he was harvesting conversations for weeks, or much longer, with some sort of filter for certain key words or phrases. This is only an assumption so far.

"More mysteriously, on the day of the attack, he somehow blocked calls to certain key responders. The caller in the control room heard the internal ringing sounds, but nothing was actually happening at the receiving end. We have no idea how he accomplished this particular sorcery." She looked at each of them in turn.

"What he inserted into the code of the reactor proper was elegant, as much as I hate to use that word. We hypothesize that a 'ringmaster' in the software directed all the other components to disperse as smaller groups of information packets, too small to trigger any automatic warning. In fact, this event was not random. Some appeared to be laid inside the internal security system itself. This is only hypothetical, but then the ringmaster may have finished the task by deleting itself."

Abbey perked up her ears. "Amazing . . . amazing . . . the parallel I'd draw is to the early days of HIV when researchers weren't initially looking for retroviruses: the kind that start as RNA, then copy as DNA to insert into our own nuclei. What bewildered the early scientists was that T and B lymphocytes were the dominant cellular targets of this retrovirus, each a central pillar of the immune system. It was like a gang coming into a new city and striking police stations instead of banks and businesses. Unbelievable."

Emmanuelle poured herself more cider. *Important to stay hydrated in the wintertime*, Emmanuelle thought. *This young man must have ideas ping-ponging around in his head.*

"I don't understand your biology, but it sounds like a possible analogy," Emmanuelle said.

"I don't understand your cryptology," Abbey said. "So we're even."

Emmanuelle nodded, leaning forward with her elbows on her knees. "It's been a long day, already. We are still uncertain how he burrowed through the firewall to the operational technology systems controlling the reactors. I will simplify to say that his attack was apparently three-pronged, and most of it was directed at weaker points external to the reactor building.

"Reactors produce colossal amounts of heat. This heat is used to produce steam to rotate the turbines that generate electricity. But the flow of water through the reactor core is also for cooling. Outside the containment building, a heat exchanger takes off the heat and returns the much cooler water back inside.

"Even if fission is halted—especially if the reactor has been operating for many months—the residual heat will create havoc. If the flow of water is halted completely, the core will melt down. If outside power is cut off, then the backup diesels need to keep the water circulating or it's a disaster.

"Jake, you know about SCADA devices, right?"

"Stands for station . . . no . . . supervisory control and data acquisition, of course," Jake said.

"Well, he managed to simultaneously hit both reactors with an attack on the heat exchangers that lie outside the reactor enclosures. The sensing circuits were manipulated to convey seemingly normal but actually incorrect information on flow rates and pressures. In fact, he was remotely opening not main but secondary valves and starting to drain the water out."

"Ye gods and little fishes." Jake paled as he seemed to consider the consequences. "But wouldn't the backup emergency core cooling system kick in?"

"It should have, but the first tier of the ECCS failed, so the operators activated the second level in the system to circumvent that transition happening automatically. The reactors had already been scrammed. Our preliminary assessment is that his code may have degraded the ECCS, but we won't know for sure until we complete a forensic deconstruction."

"Both reactors dropped out of the grid," Jake said with a grimace.

"Which brings me to the third big problem, which is exterior to the buildings. Three main high-voltage lines couple to the grid. We don't yet know how, but somehow he was able to insert multiple multiform electrical transients into these lines so that the PJM Interconnection—the responsible regional system operator—isolated and deenergized them for protection."

"And the backup generators for power?" Jake said.

"Each reactor was designed for on-site power with four dedicated diesel generators. One reactor got all four diesels online quickly, but for the second one, only one diesel booted up—as three failed—though we don't really think his software had a role in that."

"What about the heat exchangers?"

"It took at least twenty-six minutes to override the false commands to those crucial valves, with manual assistance from an outside team to confirm they were completely seated and sealed, and almost as much time to replace the missing water. Several people were badly burned by the superheated steam.

"All this was happening while security personnel were deployed to the inner perimeter fence, the roof, and other buildings all over the nuclear park. DC police were patrolling all the adjacent streets. Fortunately, they detected no concurrent physical intrusion. No kinetic attack, either."

"Kinetic?" Abbey said.

"No missile, drone, or other weapon," Emmanuelle replied.

Jake raised one eyebrow. "How can you be certain no other country or other group carried this out?"

"We can't. Not yet."

"And how do you know this bad code is not sequestered inside other reactors in the United States right now?"

"We don't know. Not yet."

"And other countries may be targets." They looked at each other for several seconds, then Emmanuelle sighed.

Jake just wouldn't stop. "Or maybe he sold this to other bad actors."

"Our preliminary psychological assessment is he wouldn't have wanted to do that."

She could almost see his next question coming. "Was there any bureaucratic bungling that delayed his identification?"

Emmanuelle had an edge in her voice this time, and knew lines showed in her face when she was tired. "You have no idea how many threats we are constantly facing." She pushed her teacup away from her and stood. "I think we're done here. I have something to give you, because I anticipated you would, in fact, accept our offer." She reached into her purse and pulled out a phone. "This looks like an ordinary cell phone, but it's encrypted. You can call the three dedicated numbers from this device listed in its directory. Each is linked to a different office.

"Don't carry this phone on your person. Keep it in the locked—remember, locked—file cabinet and keep it turned off. If you were to ever come under surveillance, this phone would not be traceable. If you ever need to contact us urgently, use only this phone—never an email or your regular phone—and identify yourself by name. No need to mention me, although one of the numbers is to my office."

"Why would we need to ever contact you urgently?" Jake asked.

"The future is hard to predict. For now, just use your imagination. We'll be calling you on your regular cell to set up appointments later. And I will remind you to get the name of that attorney that got you thinking. Bring it with you when you come in to see us, so he can be checked out. We'll probably have no direct contact with him. And even should that prove necessary, we know a fair amount about confidentiality." She watched Abbey rub her belly again.

"Abbey, when are you due?" she said.

"Next month, the fifteenth."

"The ides of March, then." Seeing the blank looks on their faces, she added, "It's Shakespearean, as in his play, *Julius Caesar*, remember?"

There was immediate comprehension on Abbey's face, but embarrassment lit up Jake's. "Gee willikers, ma'am, if that don't beat all."

Emmanuelle and Abbey both smiled, but only his honey rolled her eyes.

Emmanuelle kept her focus on Abbey. "Are you expecting a girl or a boy?"

"Yes," Abbey said, but then explained. "Jake doesn't want to know before the birth."

"Ah, it all becomes clear. Interesting." She narrowed her eyes. "I see you know how to keep a secret too." Emmanuelle gave her a slow smile. "We'll be in contact in the next few days. Thanks for the cider. I'll let myself out."

Within a minute, their mysterious visitor was gone.

"I'm feeling a bit shell-shocked," Jake said.

"You and me both," Abbey said, wide-eyed. "I'm thinking we can't discuss this with anybody except my father and mother when they get here."

Emmanuelle sat in her cold car, eyeing the tidy exterior of Jake and Abbey's inn and waiting for her windshield to defrost. *Thankfully, Jake doesn't know his cell technology well enough to know that phone I gave them was just a burner phone*, she thought, *not encrypted.*

She'd learned that using words like "encrypted," however, built trust while helping to ensure they got a commitment.

But the mystery of a boy versus a girl? I'm suddenly curious.

No need to know, not really. Not in any official sense.

Remember what happened to that cat.

THIRTY-TWO

Celebration of Hope

Never lose hope. Storms make people stronger and never last forever.
—*Roy T. Bennett*, The Light in the Heart

*Children's fables do not endure simply to establish the existence of dragons.
Of this all children are already well aware. More to the point, such tales
reassure children that dragons may be slain.*
—*Anonymous*

It was pitch-black outside. But before getting into the car, they had to stop and breathe through another contraction. When he quickly glanced up, Jake was entranced by the stars on this most momentous of moonless nights, but also appreciated a rare storm. An electrical storm, one without lightning or thunder, but truly awesome. Jake had experienced this in Minnesota, but Abbey had never seen anything like this in her life. He held her close as she hung on his shoulders for support, breathing the slow, deep breaths as she'd been taught to do.

The solar storm had brought southward a scintillating green aurora borealis, on this, the day he sensed would mark their first child's arrival, a memory that would last a lifetime, a story Jake knew would get passed down through the generations to follow.

There was a sudden gush of fluid that flowed from between her legs. "Oh . . . oh . . . my water just broke. Go back . . . get some towels, lots of them, for the car seat. Why didn't we think of this before—"

He made his voice as calming as he could. "Because we've never done this before. Okay, the door is open, you hold onto the car and the door, but don't get in yet," he said, holding his hands on her shoulders. "I'll run and get the towels, honey . . . I'll be right back." And run he did, the aurora now eclipsed by the excitement and certainty of rupture of membranes.

He took a full twenty-five minutes to drive them to the med center, anxious, but not wanting to risk an accident. The closer they got to the center of DC, the more traffic they encountered. With each contraction, Abbey grabbed his right thigh and squeezed hard. And with each squeeze, he pressed down on the gas pedal.

"She told me to come right in when I couldn't talk through a contraction on the phone with her . . . I'm so glad Charly is on call to—" Abbey stopped to breathe for a contraction starting. As it seemed to ease off, she told him, "After all the sound and fury seven weeks ago from a simple needlestick, now we get to have my midwife assist after all."

He held her hand on a straight stretch of road. "You made it to thirty-nine weeks, my dear, and somehow this was downgraded, like a hurricane, to a lower-risk pregnancy. Just a tropical storm, it seems."

"But my parents won't get here in time," she said. "They—" She was panting, then almost screeching, "will arrive three days after this kid." Jake could see a sheen of sweat on her forehead as the med center lights came into view.

"Oh, my back hurts so much . . . so much pressure."

Jake got out at the labor and delivery reception bay and practically jogged around to her side of the car. "Now, lady," he said as he opened the passenger door, "wait while I let them know we're here. I'll be right back."

He knew she could see him getting the attention of someone at the front desk. A staff person there phoned to request a wheelchair—pronto. A nurse suddenly appeared, pushing an empty wheelchair toward him.

Wheeled ingo a labor room, Abbey was then assisted in transferring to a bed covered with overlapping absorbent blue chucks. Jake stashed their carryalls in the corner and came right back to her side.

"My name is Josie, and I'm going to be your nurse all the way through. So glad to meet you both." She looked first at Abbey, then at Jake, knowing this was a first pregnancy for them both.

"I know this is intense, but just breathe through this one. Then we'll get some simple monitoring equipment in place. Good, Jake—you help her with her breathing by mirroring what she's doing."

He figured she was partly trying to keep him occupied.

"I'm going to put a blood pressure cuff on your right arm and this will stay on unless we have you walking around later. It will tighten up from time to time. Just keep your arm muscles relaxed. Let's listen to the baby's heart now." She applied lubricant to Abbey's left lower abdomen, and soon everybody could hear the heartbeat as the pain started in again.

Listening through the entire contraction, Jake sensed the nurse was satisfied with the pattern.

"Good, good, that one's over," she said. "Now, Jake, you help her lift up so we can get this webbing around her abdomen."

"Just pull it . . . up. Like this . . . ?" Jake asked.

"You got it. Now, I'm going to collect a sample of vaginal secretions for another PCR, which you knew about."

All three appeared sober and serious for the few seconds this took.

"Next, I need to check her," she said, pulling on a glove. "Abbey, I'm going to have you hold your legs apart . . . that's right." Josie had one hand on Abbey's thigh, the other reached toward Abbey's introitus. "You'll feel some pressure on your perineum down toward the bed, then I'm going to feel the position of your baby's head. Try not to tense up. This is just like the exams you've had in the office. Okay, baby is head down, in what we call a left occiput anterior position. You're already four centimeters dilated with the head almost down to zero station. Well-engaged in the pelvis, if you remember that from school."

Abbey took a deep, cleansing breath after the contraction and said, "I remember that from med school, LOA position."

Josie pulled the webbing down from Abbey's upper abdomen. Then she slid the two monitors underneath, one for contractions and the other for fetal heart tones. The regular beat filled the room.

"Oh, good . . . when does Charlene come in? She's the only one I've seen in the whole pregnancy except when I was in the hosp—oh, wait, here's another one." Abbey gasped, her panting getting louder at the end. Josie cleansed her flushed forehead with a cool, wet washcloth.

"Breathe, breathe. Good. Well, I know everyone in L&D has been anticipating your arrival for weeks, as you can imagine. Once the baby is out, Charly will collect the usual cord-blood samples, but also another PCR. Bet you're tired of hearing that particular term, eh?"

"I thought I detected a bit of a Canadian accent," Jake said with a chuckle, and Josie lifted her chin and gave him a half-smile.

"Charly usually waits until primips are about six centimeters before she comes in."

"Jake," Abbey said, rotating in bed. "A primip is a first, full-term pregnant woman, short for 'primiparous' . . . wait, that's not right, a woman pregnant for the first time is a primigravida, right, not yet delivered. A primip is someone who's already had one birth. That's a bit confusing, isn't it?" She deepened her breathing for the next contraction.

"You're right, Dr. London. But by long-held convention, it's just faster to say 'primip.' Listen, you're all hooked up on the monitors. I need to get this sample to the lab and some ice chips for you," Josie said. "Do you think you're going to want some medication?"

Abbey grimaced and shook her head.

"One last thing: Please finish rolling to your left side. You may remember how that improves placental blood flow. Plus, it's more comfortable." Rubbing her chin with the back of her hand, she said, "Jake, if she has back pain, massage or put pressure on her lower back. That often helps.

"If you'd like, I could send you two out to walk the hall. Those monitors transmit wirelessly." She smiled on her way out. "Remember, when the time comes that you feel an urge to push, I need to know right away. You know where the call button is."

The two lovebirds slowly walked the halls, whispering to each other, stopping with each contraction for her to hang half her weight on his. He was her rock and he knew it, even if he had doubts inside. *And I certainly know how strong she is.*

Nurses smiled encouragement as they walked by in either direction.

Not long after they got back to the room, in walked Charly. "Hey, you two, how're you doin'?" She crossed the room with an air of quiet command. "I looked at the monitoring data and it seems your baby is tolerating this just splendidly."

"Abbey is only sorry her parents will miss the show," Jake said. "I have been entrusted to keep them updated."

Charly smiled. "Sounds like something along the lines of sending men out to boil water and get clean sheets to keep them out of the way."

"Imperfect analogy at best," Jake said. "Besides, I know we're not supposed to do video, but we did a quick video call with her folks."

"So, they're excited?"

"Yes, a first grandchild for them. So yes, definitely."

"And you also understand the exhaustion you're experiencing now may continue for months with a newborn at home," she said softly. Jake realized she probably got to use these lines on every patient she admitted.

Jake broached a question. "Charly, I know you know whether this is a boy or girl."

"Of course, we have access to all the sonographic data."

"Isn't it close enough now that you can tell me?" He tried his best rueful smile.

She laughed, saying, "I know what you're trying to do and it's not going to work. You'll find out soon enough. In fact, this would be a good time to check on Abbey's progress."

She leaned over to check on Abbey's progress and a grin took over her face. "I think you're trying to steal a march on us—you're eight centimeters and the head is down to a plus-one station. Are you feeling like you need to push at all?"

"A little, but I know I'm not supposed to yet."

"Soon. We'll keep an eye on you." She paused. "You could try walking more if that appeals to you. It might speed things along."

Abbey looked at Jake.

"Whatever you want, babe," he said.

"Let's do it. You two help me up before the next one hits."

But she and Jake had only traversed the corridor once before she really sagged in his arms and began giving in to the irresistible urge to bear down. Jake called out and Josie and Charly helped them back into the room.

Charly examined her again. "Are you sure you haven't done this before? You've only been in labor eight hours as a primip, and here we are. You can see where my hand is, and your baby is trying to push it back out. Time to push away. Jake, you help move her shoulders forward so she can grab her knees . . . yes, yes, just like that."

He felt himself pushing with her, reminded that he needed to empty his bladder soon.

The conversation stilled between contractions. Josie persuaded Abbey to go limp as a rag until the next contraction, which came quickly. The head was beginning to show. Jake had washed his hands for this moment. He asked permission before reaching down and touching the slippery crown of his baby's head.

"Blood is blood, mucus is mucus, but what's all this white stuff?" he asked in a hoarse whisper.

"Vernix—vernix caseosa, formally. Part of many deliveries. Helped to protect the baby's skin during all those months in the amniotic fluid."

Jake thought of all the times he had touched those tissues gently, and now this baby was stretching it like a big O-ring.

He saw Abbey entering another contraction, so he moved back to tighten his grip on Abbey's hand—if only to protect his fingers from getting crushed in return. He remembered what people had said about the strength of women in labor.

Abbey inhaled deeply, then pushed, pushed to beat the band. Abbey began with a grunt, then pulled on her legs and let out a long, low moan, while Charly eased the tissue around the baby's head. "Don't push now, Abbey, don't push. Boy—this kid has a real headful of hair." Quickly, she suctioned the baby's nose and mouth. Then, with a rush of amniotic fluid and blood, the baby was out and laid on its side. Charly toweled the baby's face from top to bottom, further suctioning the nose and mouth, then vigorously rubbed its back to stimulate breathing.

"Come on baby, now's the time to cry," she said, and the girl let out a lusty howl, pulling in breaths to cry some more. Jake was exploring her arms and legs and counting fingers and toes, trying not to cry tears of relief.

He gave Abbey a big hug. "You did it, honey!"

"You've got a beautiful daughter," Charly said. "And Josie is going to put her right on your bare chest. So pull up your gown and she'll cut off the webbing and take away the monitors. That's the way. Hold her with both hands. Jake and Josie will help wipe her down further.

"Josie, when you're done massaging the fundus, I'm going to need a surgical tray with some lidocaine and three-ought vicryl suture . . . Abbey, you had a bit of tearing of the perineum, a small second-degree injury that should heal up easily.

"Very normal newborn, Josie," Charly said. "How about Apgars of seven and nine? Jake, those are just the scores of neonatal vitality in the first five minutes of life."

He just shrugged. *I know my daughter's fine. I can tell.*

"Abbey, since I'm sure you know all the anatomy here, I'll tell you the short left periurethral laceration does not require repair, though it will burn a bit when you pee for a couple of days. You delivered at thirty-nine weeks with a medium-sized baby for that gestational age, which is why it went pretty quickly."

Charly clamped the umbilical cord and surprised Jake by handing him, without forewarning, the handle of an extra pair of scissors to cut between the two clamps.

The nurse had the injection ready—insurance against resurgent bleeding—and gave it into the upper outer gluteal area.

The room grew quiet now that the excitement was winding down.

After Charly and Josie successfully removed the placenta, the room seemed to get quieter now that the excitement was winding down. Jake took a turn holding his swaddled daughter, who turned her head to either side and breathed tiny breaths.

Josie adjusted the light on the perineum, while Charly injected local anesthetic and began suturing. "Abbey, I know you know this, but you've been up all night. Josie may want to get you an ice pack to put on this area, and she'll be checking the bleeding and massaging your uterus periodically. Though it's a good idea to start massaging it yourself when you see her coming. These nurses lift weights, I tell you, and you'll be having some afterpains anyway."

"Would you two like some lunch?" Josie said. "I can put in an order for you."

Charly stood up and stretched her back again. "Speaking of orders, I need to get my notes done before I head to the office. One last question: What is this darling's name?"

Abbey looked deeply into Jake's eyes, and he looked back without saying a word . . . waiting . . . questioning.

"I was in such a hellhole seven weeks ago. I told myself if I survived, we had to name this baby . . . 'Hope,'" Abbey said. "Because that's all I had."

He nodded. He tried to speak and failed. He squeezed her arm, and finally said, "I love it. Kind of an older name, but pretty. Let us have . . . Hope."

"And how about Aurora as a middle name?" Jake said after some thought.

Abbey's smile rose as slowly as the Moon. "I like that, Jacob. So let me propose . . . Hope Aurora Harper."

He looked into his wife's resolute eyes. "We will tell many tales to this child, my love. Of dragons and princesses. Of heroes and villains. Of how this day came to be."

Anne's and Colby's eyes had both sparkled when they saw the phone video of the family that now numbered three. Anne had teared up, promising they'd be there in a couple of days.

The early efforts at breastfeeding were gong well, Hope demonstrating strong root and suck reflexes.

Josie came in with a package and a card. "This just got dropped off at the front desk, sort of a special delivery. Lady didn't ask if she could see you, and as you know, hospital policy states it's too early for non-family visitors." She left the box on the bedside table and went to check on another patient.

Jake opened the envelope as the new mom was busy trying to get their baby to latch on. "It's from Emmanuelle, and the card has a quotation from somebody named Nicole Lyons." He handed it to Abbey.

She adjusted Hope's position against her left breast and peered at the words on the card: "She's an old soul with young eyes, a vintage heart, and a beautiful mind."

Abbey looked at Jake, puzzled. "She couldn't have known, could she? Open the package and let's see what she got us."

He tore off the gift wrap, opened the box, and peeled back the tissue paper. "It's what I think is called a jumper. But will you get a look at this," he said as he was pulling it out.

A jumper in beautiful light pink, with darker mauve piping as trim.

One Journey Ends

What experience and history teach is this—that people and governments never have learned anything from history, or acted on principles deduced from it.
—*G. W. Friedrich Hegel*

As I stand mute under the north porch, the Big Dipper reminds me of the still, chill beauty of the night sky, slowly cartwheeling as it has for billions of years and will continue for billions more—the voiceless chorus of the spheres. Early dawns the morning, but now it is earlier yet. At the cusp between its third and fourth quarters, a waning crème brûlée moon casts light from the southeast. No barn owl espied this morning, though I occasionally see one lofting its way home to roost.

Returning to my study, my hands cradle the coffee mug, not drinking but just inhaling its sustenance and enjoying its warmth. The ticking of the old clock on the wall is companionable but a reminder, second by second, of the inexorable passage of days. Sunflower petals sashay across our patio under a frolicsome breeze, whispering through our thick walls. We've witnessed a year of praying mantises, more than we've ever seen, and I wonder where they are sheltering now, at the beginning of autumn. We all need shelter from the coming storms.

My characters are speaking to me, struggling to get their words out. What I hear them whispering is that their lives, their journeys, are not yet complete. They have gotten more real to me as time has passed, almost breaking through the veil. Daily, I have struggled to absorb the craft of writing. Apparently that eighth grade creative writing class, as excellent as my teacher was, does not constitute a sufficient education. Writing is strenuous work, as difficult as trying to throw away a boomerang. And perhaps as dangerous.

Jake and Abbey have bustling lives, now made more challenging by parenthood. How would they ever find the time to march through another chronicle, another narrative, another book? In the same way we all manage the innumerable tasks of our lives. The realization creeps up on me, now a certitude, that I do want to see where their lives lead, how all this plays out. I am an obstinate optimist, and these two as well exhibit a lot of resilience, a lot of spunk. How will young Hope turn out? Only time will tell. How will our hope for the future play out? That . . . that is the question of our age.

Climate change is a Shakespearean tempest, a storm surge the planet has not seen for thousands of years, the mother of all calamities as far as our civilization is concerned.

Climate change is a fearsome dragon.

Gird yourself for the battle.

Acknowledgments

I cannot pretend I am without fear. But my predominant feeling is one of gratitude. I have loved and been loved; I have been given much and I have given something in return; I have read and traveled and thought and written. I have had an intercourse with the world, the special intercourse of writers and readers.
—Oliver Sacks

If each of my words were a drop of water, you would see through them and glimpse what I feel: gratitude, acknowledgement.
—Octavio Paz

In terms of the writing process, I would first and foremost mention my multiple teachers and editors. Sydney Strauss reviewed first chapters of earlier drafts before being called away, Kiffer Brown and David Beaumier and the other good folks at the Roost at Chanticleer provided support and encouragement, and Scott Taylor reviewed the entire manuscript, as he did for my first foray into writing. Savannah Gilbo reviewed a later draft, Cami Ostman and Lisa Dailey assisted at a Wayfaring Writers' conference, and finally, Britta Jensen brought in the anchor leg with the final draft. Other than educators and editors, only a couple of people perused every single word, but here, I gratefully mention Rich Gelinas and Bob Good.

Our son, Yancey Lawrence, an experienced emergency medicine physician, reviewed chapters three and five for medical verisimilitude, and suggested several crucial corrections.

For nuclear power central to chapters six and eight, Nathan Wilson, an electrical and software engineer, committed himself to a lengthy discourse over nuclear energy—exceptionally lively because we found ourselves diametrically opposed regarding the future of this industry, both domestically and globally. No doubt we are both looking forward to seeing where the future takes us.

My friend Cliff Halvorsen, catfish farmer and AP high school science teacher, generously offered to review what finally evolved into chapters twenty and twenty-two, with a particular emphasis making them less recondite and more approachable.

Doug Clark, an associate professor in the geology department at Western Washington University, has a primary focus on the geomorphic and paleoclimatic character of alpine glaciation, with a particular interest in the later Pleistocene and Holocene epochs. He certainly picked a great place in Northwest Washington to study these matters. And he was extraordinarily helpful in helping me update chapters eleven and thirteen, the first examination of climate science in this book.

Dan Welch, an architect of my acquaintance and expert in efficient home design, provided a professional overview of chapter twenty-three. He also deserves significant credit for our certified Passivhaus™ home, which we have found to be our most comfortable residence ever.

Other friends who participated as beta readers or in discussion of this book include, in alphabetical order, Sara Devlin, Clare Fogelsong, Faye Hayes, Dave Irwin, Roger Kerlin, Duane Meares, and Doug Smith. You know who your real friends are—the ones who are willing to read and comment on a manuscript.

No less important were the group of beta readers in a virtual Village Books nonfiction writers' group in Bellingham, whose sage advice and resilience through several years of the coronavirus epidemic helped me immensely. In alphabetical order, they include, P. J. Beaven, Dave Blander, Elaine Brent, Olivia Coleman, Kendra Cook, Katie Fleischmann, Dick Gerry, Larry Guevara, Heidi Hull, David Hustoft, Nicki Lang, Carol Lawrence (no relation), Donna LeClair, Verna May, Melody Rhode, Tiahna Skye, and Rod Spencer.

I owe special recognition as well to the members of the in-person Village Books fiction writers' group in Bellingham, whose prescient observations and often scholarly advice served me very well. In alphabetical order, they include, Tom Altreuter, Lee A. Brown, Paul Clinton, Alissa DeLaFuente, Jayne Entwistle, Joe Evergreen, Brian Feutz, Joe Henderson, Jacqueline Martin, Kenneth Meyer, Evan Mielke, Robert S. Phillips, Gail Pinto, Larry Scherer, Matt C. Scott, Al Skinner, Meghan Squires, Monica Woelfel, Bob Zaslow, and Geoff Ziezulewicz.

The estimable artist Bob Paltrow receives credit for the cover of both this and the next book, and for talking me through multiple iterations. In this, he was assisted by the dragon motif, the beautiful and mythical Prismatic Dragon Head, which I was fortunate to find in Wikipedia Creative Commons, licensed with public domain dedication under the category CCO 1.0 Universal, such that the person who is associated with this deed has dedicated his or her work to the public domain by waiving all rights to the work worldwide under copyright law, including all related and neighboring rights, to the extent allowed by law.

As a disclaimer, aside from an online thesaurus, no artificial intelligence was involved in the researching, writing, or editing of this work.

Finally, I note the District of Columbia shares borders with Maryland and Virginia, and connects with lands along the Anacostia and Potomac Rivers. I also recognize the historic presence and influence, dating back at least four thousand years, of the Piscataway, Pamunkey, Nentego (Nanichoke), Mattaponi, Chickahominy, Monacan, and Powhatan Native Americans. I formally acknowledge their sovereignty and stewardship as Indigenous peoples, who have an enduring relationship with these traditional territories.

Notes

1. "Pegasus (Spyware)." *Encyclopædia Britannica*, August 21, 2023. https://www.britannica.com/topic/Pegasus-spyware

2. "FBI Confirms It Obtained NSO's Pegasus Spyware." *The Guardian*, February 2, 2022. https://www.theguardian.com/news/2022/feb/02/fbi-confirms-it-obtained-nsos-pegasus-spyware/

3. "Elliptic-Curve Cryptography." Wikipedia, September 1, 2023. https://en.wikipedia.org/wiki/Elliptic-curve_cryptography

4. "Nerds, Ninjas, and Neutrons: The Story of the Nuclear Emergency Support Team." *Bulletin of the Atomic Scientists*, July 6, 2023. https://thebulletin.org/premium/2023-03/nerds-ninjas-and-neutrons-the-story-of-the-nuclear-emergency-support-team/

5. "Historic Nuclear Accident Dashed Swiss Atomic Dreams." SWI swissinfo.ch, January 21, 2019. https://www.swissinfo.ch/eng/multimedia/radioactive_historic-nuclear-accident-dashed-swiss-atomic-dreams/44696398

6. "Tokaimura Nuclear Accident." Nuclear Energy, May 10, 2010. https://nuclear-energy.net/nuclear-accidents/tokaimura

7. World Nuclear Industry Status Report, 2020. https://www.worldnuclearreport.org/IMG/pdf/wnisr2020-v2_hr.pdf

8. Mervine, Evelyn. "Nature's Nuclear Reactors: The 2-Billion-Year-Old Natural Fission Reactors in Gabon, Western Africa." Scientific American Blog Network, July 13, 2011. https://blogs.scientificamerican.com/guest-blog/natures-nuclear-reactors-the-2-billion-year-old-natural-fission-reactors-in-gabon-western-africa/

9. "Neutron Moderator." Energy Education. Accessed October 19, 2023. https://energyeducation.ca/encyclopedia/Neutron_moderator

10. "Nuclear Transmutation." Socratic.org. Accessed October 19, 2023. https://socratic.org/chemistry/nuclear-chemistry/nuclear-transmutation

11. Valjak, Domagoj. "In 1980, Chemist Glenn Seaborg Solved a Centuries-Old Problem in Alchemy and Turned a Non-Precious Metal into Gold: The Vintage News." The Vintage News, December 6, 2017. https://www.thevintagenews.com/2017/12/06/glenn-seaborg-alchemimist-in-1980/

12. Collins, Allen, Ben M. Waggoner, and David Smith. "Introduction to Cnidaria: Jellyfish, Corals, and Other Stingers." UC Museum of Paleontology, June 19, 1994. https://ucmp.berkeley.edu/cnidaria/cnidaria.html

13. SLR Document Changes: Add FE Section 3.5.2.2.2.7 and AMR Items for Irradiation Embrittlement of Reactor Vessel (RV) Steel Supports and Other Steel

Structural Support Components near RV, n.d.
https://Adamswebsearch2.Nrc.Gov/webSearch2/Main.Jsp?AccessionNumber=
ML20049H359

14. Lochbaum, Dave. "The Bathtub Curve, Nuclear Safety, and Run-to-Failure." The
Equation, November 17, 2015. https://allthingsnuclear.org/dlochbaum/
the-bathtub-curve-nuclear-safety-and-run-to-failure

15. "Tropical Rainforest Primary Productivity." Hugo Ricci. Accessed October 30,
2023. https://tropicalrainforestbiomebyhugo.weebly.com/
primary-productivity.html

16. Shahzaib. *Smart Land Use: Palm Oil Is The World's Most Efficient Oil Crop.* Malaysian
Palm Oil Council. https://mpoc.eu/smart-land-use-palm-oil-is-the-worlds-most-
efficient-oil-crop/

17. U.S.Transportation Sector Greenhouse Gas Emissions 1990 –2018 , n.d.
https://nepis.epa.gov/Exe/ZyPDF.cgi?Dockey=P100ZK4P.pdf

18. Tcs. "Federal Subsidies for Corn Ethanol and Other Corn-Based Biofuels." Taxpayers for Common Sense, April 15, 2022. https://www.taxpayer.net/energy-natural-
resources/federal-subsidies-for-corn-ethanol-and-other-corn-based-biofuels/

19. *Algae for biofuel production.* Farm Energy. April 12, 2019. https://farm-energy.
extension.org/algae-for-biofuel-production/

20. The Trustees of Princeton University. (n.d.). *José Caraballo-Cueto | Program in Latin
American studies.* Princeton University. https://plas.princeton.edu/people/
jos%C3%A9-caraballo-cueto

21. Smith, T. J., & Banks, D. W. (2022, August 13). *Can the F-150 lightning make everyone
want a truck that plugs in?.* The New York Times. https://www.nytimes.com/2022
/08/13/business/electric-vehicles-battery-factory-georgia.html

22. Niedermeyer, E. (2022, August 27). *You want an electric car with a 300-mile range? when
was the last time you drove 300 miles?.* The New York Times. https://www.nytimes.com
/2022/08/27/opinion/electric-car-battery-range.html

23. *Circular supply chain for lithium-ion batteries.* Redwood Materials. (n.d.).
https://www.redwoodmaterials.com/

24. Wikimedia Foundation. (2023, May 23). *General mining act of 1872.* Wikipedia.
https://en.wikipedia.org/wiki/General_Mining_Act_of_1872

25. Davenport, C., Friedman, L., & Plumer, B. (2022, August 24). *California to ban the sale
of New Gasoline Cars.* The New York Times. https://www.nytimes.com/2022/08/24
/climate/california-gas-cars-emissions.html

26. "Layers of Earth's Atmosphere." UCAR Center for Science. Accessed October 31,
2023. https://scied.ucar.edu/learning-zone/atmosphere/layers-earths-atmosphere

27. *Karakoram.* Infogalactic. (n.d.). https://infogalactic.com/info/Karakoram

28. U.S. National Library of Medicine. (n.d.). *Dimethyl sulfide.* National Center for Bio-
technology Information. PubChem Compound Database.
https://pubchem.ncbi.nlm.nih.gov/compound/Dimethyl-sulfide

29. *Wet deposition.* ScienceDirect. (n.d.). https://www.sciencedirect.com/topics/earth-
and-planetary-sciences/wet-deposition

30. U.S. Geological Survey. (n.d.). *Lahars of Mount Pinatubo, Philippines.* Lahars of Mount
Pinatubo, Philippines, fact sheet 114-97. https://pubs.usgs.gov/fs/1997/fs114-97/

31. Salopek, P. (2023, May 18). Inside the deadly world of India's Sand Mining Mafia. https://www.nationalgeographic.com/environment/2019/06/inside-india-sand-mining-mafia/

32. Centers for Disease Control and Prevention. (2022, August 5). *Rickettsia parkeri rickettsiosis*. Centers for Disease Control and Prevention. https://www.cdc.gov/ticks/tickbornediseases/rickettsiosis.html

33. Mandavilli, A. (2023, February 15). *How climate change is spreading malaria in Africa*. The New York Times. https://www.nytimes.com/2023/02/14/health/malaria-mosquitoes-climate-change.html?action=click&module=Well&pgtype=Homepage&section=Health

34. Schumacher, R. S., & Rasmussen, K. L. (2020, June 2). *The formation, character and changing nature of mesoscale convective systems*. Nature News. https://www.nature.com/articles/s43017-020-0057-7

35. Jet Propulsion Laboratory. (n.d.). *Oceans Melting Greenland*. NASA. https://omg.jpl.nasa.gov/portal/

36. ScienceDaily. (2017, October 4). *In warmer climates, Greenlandic deltas have grown*. ScienceDaily. https://www.sciencedaily.com/releases/2017/10/171004133554.htm

37. Cook, J., Takeuchi, N., Irvine-Fynn, T., & Edwards, A. (n.d.). Cryoconite: The dark biological secret of the Cryosphere. https://journals.sagepub.com/doi/10.1177/0309133315616574

38. Wikimedia Foundation. (2021, December 9). *Bentley Subglacial trench*. Wikipedia. https://en.wikipedia.org/wiki/Bentley_Subglacial_Trench

39. X, S. (2020, January 28). *Scientists drill for first time on remote Antarctic Glacier*. Phys.org. https://phys.org/news/2020-01-scientists-drill-remote-antarctic-glacier.html

40. Mooney, C. (2019, February 8). *Earth is "missing" at least 20 ft of sea level rise. Antarctica could be the time bomb*. ScienceAlert. https://www.sciencealert.com/earth-s-climate-s-now-like-115-000-years-ago-when-the-sea-was-much-higher

41. Hansen, J., et al. "Ice melt, sea level rise and superstorms: evidence from paleoclimate data, climate modeling, and modern observations that 2°C global warming is highly dangerous." https://bpb-us-e2.wpmucdn.com/faculty.sites.uci.edu/dist/b/662/files/2017/06/Ice-melt-sea-level-rise-and-superstorms-evidence-from-paleoclimate-data-climate-modeling-and-modern-observations-that-2C-global-warming-is-highly-dangerous.pdf

42. *Der Ring des Nibelungen*. Hessisches Staatstheater wiesbaden. (n.d.). https://english.staatstheater-wiesbaden.de/opera/ring-cycle/

43. *Tracking the costs of the corrosion epidemic*. watermainbreakclock.com. (n.d.). https://www.watermainbreakclock.com/

44. Zarza, Laura F. "World Toilet Day 2019: Leaving No One Behind." Smart Water Magazine, November 22, 2019. https://smartwatermagazine.com/blogs/laura-f-zarza/world-toilet-day-2019-leaving-no-one-behind.

45. Wilson, C. (n.d.). *Florida's water: A truly septic problem*. Dewberry. https://www.dewberry.com/news/blog/post/blog/2016/10/06/florida's-water-a-truly-septic-problem

46. National Science Foundation. (2023, December 23). *Toward a Cenozoic history of atmospheric CO_2 | Science*. https://www.science.org/doi/10.1126/science.adi5177

47. Armstrong McKay, et al. "Exceeding 1.5°C global warming could trigger multiple climate tipping points." *Science*. https://www.science.org/doi/10.1126/science.abn7950

48. Sherwood, S.C., et al. *An assessment of Earth's climate sensitivity using multiple lines of evidence.* Advancing Earth and Space Sciences. https://agupubs.onlinelibrary.wiley.com/doi/epdf/10.1029/2019RG000678

49. *Climate model: Temperature change (RCP 2.6) - 2006 - 2100.* Science On a Sphere. (2013, November 17). https://sos.noaa.gov/datasets/climate-model-temperature-change-rcp-26-2006-2100/

50. NOAA . (n.d.). *Climate Forcing.* Climate.gov. https://www.climate.gov/maps-data/primer/climate-forcing

51. The Kigali Amendment (2016): The amendment to the montreal protocol agreed by the Twenty-eighth meeting of the parties (Kigali, 10-15 October 2016). UN Environment Programme. (n.d.). https://ozone.unep.org/treaties/montreal-protocol/amendments/kigali-amendment-2016-amendment-montreal-protocol-agreed

52. Mingst, K. (2023, October 20). *International Civil Aviation Organization.* Encyclopedia Britannica. https://www.britannica.com/topic/International-Civil-Aviation-Organization

53. Carbon Offsetting and Reduction Scheme for International Aviation (CORSIA)." ICAO. Accessed October 31, 2023. https://www.icao.int/environmental-protection/CORSIA/pages/default.aspx

54. Dwortzan, M. (2016, May 1). *Steven Barrett: Contrails, Carbon & Climate.* MIT Global Change. https://globalchange.mit.edu/news-media/jp-news-outreach/steven-barrett-contrails-carbon-climate

55. Sengupta, S. (2020, May 29). *Economic Giants are restarting. here's what it means for climate change.* The New York Times. https://www.nytimes.com/2020/05/29/climate/coronavirus-economic-stimulus-climate.html

About the Author

Photo by Cheryl Crooks Photography

Born in Minnesota, I grew up in Missouri, then at eighteen realized my California dream with undergraduate years at Stanford, including two years on the tennis team. After med school back in my home state, I completed a residency at UC San Francisco, where I gave my first public lecture, a talk on nuclear power. Two weeks later I started teaching full time in a UC Davis-affiliated family medicine residency, with much of my practice focused on obstetrics but also HIV care starting in early 1983. I taught full time for sixteen years, with my main academic appointment at UC Davis, but coming full circle, a community faculty appointment at Stanford Med Center as well, since I taught a number of their PA students.

I lost my heart in San Francisco. My wife became a certified nurse-midwife, and between the two of us we delivered thousands of babies and had three of our own, all graduated and married—two of them with a couple of grandsons apiece. We moved up to Bellingham more than a quarter century ago, but are now both retired. In the last ten years I have continued to lecture widely, including several seminar series at Western Washington University, plus many other venues, including a couple of out-of-state conferences. But I no longer teach medicine, rather, I teach about energy systems, the climate system, epidemiology, and the electric grid, and this set of fourteen or so lectures form the basis for many of the chapters in my book.

Our biggest climate-related trip was in Europe in 2015, which included biking in Berlin, Copenhagen, and Oslo, then a two-week cruise from Svalbard to Greenland and Iceland. We accomplished hiking, kayaking, and a true polar plunge well north of the Arctic Circle while in Greenland. Additionally, we explored some of the eastern Canadian maritime islands in 2023. In the next twenty months we plan to visit and investigate the electric grids of Kauai and Puerto Rico.

Milton Keynes UK
Ingram Content Group UK Ltd.
UKHW022111190224
438095UK00017B/768